FUNDAMENTALS
OF PESTICIDES/
— A SELF-INSTRUCTION GUIDE

SECOND EDITION

George W. Ware
UNIVERSITY OF ARIZONA

THOMSON PUBLICATIONS
P.O. BOX 9335
FRESNO, CA 93791

Copyright© 1986 by George W. Ware
Printed in United States of America
Library of Congress Number 81-86070
ISBN 0-913702-35-8

DISCLAIMER CLAUSE

The author and publisher are in no way responsible for the application or use of the chemicals mentioned or described herein. They make no warranties, expressed or implied, as to the accuracy or adequacy of any of the information presented in this book, nor do they guarantee the current status of registered uses of any of the chemicals with the U.S. Environmental Protection Agency. Also, by the omission, either unintentionally or from lack of space, of certain trade names and of some of the formulated products available to the public, the author is not endorsing those companies whose brand names or products are listed.

PREFACE

The general public knows virtually nothing about pesticides. This may be due to the adverse influence, if not sensitization, by the news media, to the undesirable features of pesticides, and being alerted to clean environment, the ecology, cancer risk, birth defects and other subjects, usually from exaggerated or misrepresented points of view. Today's news reporting brings synthetic chemicals to the public's attention with sensational journalism and headlines, mixing words such as poisons, pesticides, biocides, food additives, chemicals, oil spills, water quality, environmental contamination and residues in scare-articles intended to excite and generate fear in the reader. There is a greater need now than ever before for enlightening the uninformed by providing an accurate yet readable self-teaching guide.

By studying this book you become the one person in 10,000 who is developing a balanced concept of pesticides and their importance to our everyday life. Congratulations!

Pesticides, chemicals used to control the harmful organisms we know as pests, play a significant role in the life of modern man. To hold them in disdain is to ignore the facts. We have isolated ourselves from the "back to nature" philosophy by the steady ascent of our affluence, and it would be virtually impossible to return to agricultural practices of as recently as 1970. There is simply not enough available manpower to plant, thin, cultivate, fertilize, weed, irrigate, and harvest our crops as we did then. All of this is done in the mid-1980's with powerful, man-replacing machinery, assisted by the judicious use of those "super chemicals", pesticides.

This self-instructional guide on pesticides was first conceived as a collection of separate "programmed learning" lectures presented to agricultural pest control advisors, structural pest control operators, garden clubs, and civic groups. Though everyone knew something about pesticides, their knowledge was usually fragmentary, at best, if not wrong. Nearly everyone in the audiences enjoyed learning even the most intricate details of pesticide chemistry, when the information presented was constructed upon old familiar building blocks. This was done by beginning beneath the level of understanding of the audience, by holding their attention, and by moving slowly into unfamiliar territory. It seemed then that material on pesticides could be prepared in a palatable auto-tutorial form for interested persons from varied backgrounds. This book is the result of my efforts to accomplish that feat.

My aim is to explain pesticides to the layman-to give an appreciation for the present state of this segment of chemical art and science. The entries are not restricted to a bare definition: some information about most of the pesticides named is given, so as to convey something of

their significance in late 20th century agricultural and
structural pest control technology.

In writing the second edition of this self-instruction
book, I have tried to adopt a simple, factual approach to a
subject that is all too often emotionally charged. Also,
I have made an effort to avoid the use of scientific and
technical terms prior to their introduction in the learning
units. To avoid the inherent dangers of oversimplification,
however, some compromise has been necessary between the too-
simple and the incomprehensible.

Because of the tremendous amount of material to be
covered, no book of this kind could be complete. This
guide is intended to present a comprehensive picture of the
major classes of pesticides, and not an exhaustive study or
catalog. For instance, I have not covered avicides, molluscicides, piscicides, algicides, disinfectants and other
smaller groups. Thus the exclusion of pesticides in no way
indicates an endorsement of those illustrated or discussed.
I hope this self-pacing guide will be of interest, not only
to the layman, but also to students, scientists, and especially those preparing for federal or state certification
and licensing in the field of pesticide usage.

Acknowledgements. During the revision of this text,
I have received valuable advice, criticism, and assistance
from many people. In particular, I would like to record my
gratitude to:

Ms. Susanne T. Cotty, for painstakingly reviewing the
first edition and making suggestions for the revision; W. T.
"Tommy" Thomson, for updating the units on herbicides and
fungicides; Dr. Theodore G. Tong for updating the unit on
pesticide toxicity; and to Ms. Grace Baker, who typed and
edited the manuscript with the utmost attention to detail,
as if it were her own.

Finally, to my wife, Doris, who assisted me in the
indexing and sacrificed many evenings and weekends, permitting me to complete this second edition, I owe a special
debt of appreciation.

                                          George W. Ware

Tucson, Arizona
October 1985

CONTENTS

VIII

INSTRUCTIONS FOR USING THE MANUAL

This is a do-it-yourself or autotutorial book. Its format is designed to help you use it as an aid for self-instruction. At the end of Unit 1, where the questions begin, place the answer shield on the page to cover the answers while exposing only the question. Write the answers you think are correct, then slide the shield down just enough to reveal the answer. In some sections several questions are presented on a single page.

You may be tempted to peek at the correct response before recording your answer, but this will stop when you realize that the purpose of the book is to teach rather than test. You are the only person who will be checking for errors, and no one cares how many mistakes you make as long as you are convinced at the end that you know what is correct.

The time required to work through the manual will vary with the reader's background and study habits. The average time should be about 17 hours, varying between 12 and 20 hours.

Tests with self-instruction and programmed texts have shown that students who complete the study by writing the answers score higher on tests than those who do not. So if you are to achieve the objectives which the program is designed to help you attain, you should work through the material in the sequence presented, and write your answers.

Beginning on page 243 is the Final Exam, 125 carefully selected questions which highlight the most important points of the manual. Unlike the rest of the book there are no answers provided with these. Instead, the bracketed number following each question is the page where the answer can be found. You should complete this exam before taking applicator certification or pest control advisor licensing tests.

Good luck!

LEARNING OBJECTIVES

Before digging into *Fundamentals of Pesticides*, I should tell you what it is supposed to do. All self-instruction books should have a definite study-audience in mind and some rather clear-cut goals stated in terms of the abilities or skills the student will be able to demonstrate upon completion of each unit. In our new communication age these goals become "learning objectives".

This book is aimed first at pesticide applicators, both private and commercial, agricultural pest control advisors, pesticide salespersons, and structural pest control operators, who are preparing for certification or licensing. Second, it is designed for college students as a useful supplement to materials assigned for study in courses on ecology and the environment, pest control, plant protection and agricultural chemicals. Third, it is intended to be used by state and federal officials involved in the training and preparation of materials for persons seeking licensing and certification as required by state and federal legislation. Fourth, and last, it is written for the concerned citizen, lay members of garden and civic clubs, who seek a better understanding of the science and technology guiding their lives.

Fundamentals of Pesticides was developed to teach a selected set of objectives which the student should understand and be able to demonstrate after each unit. Instructors may use them as standards for measurement, both for designing their input and for preparing tests. The student may be further motivated for study when he realizes that he has a convenient checklist of terminology and chemistry skills to be mastered instead of tediously wading from front cover to back, hoping to grasp the overall gist of the book.

If he has progressed through the material and successfully answered the accompanying questions, the student should be able to:

Unit 1 Pesticides--Chemical Tools for Pest Control

a. Comprehend the role of pests in history, particularly insects as vectors of disease.
b. Delineate the pests of agriculture and suburbia.

c. List the vast array of pesticides in use today.
d. Predict the results of withholding all pesticide use
   in the U. S. for one year.
e. Estimate the significance today of a severe outbreak of
   plant disease considered catastrophical historically.
f. Summarize the underlying reasons for not returning to
   agricultural production as practiced a generation ago.
g. Explain the total dependence of our society on pesticides for food, fiber, and protection from insect-trans-
   mitted disease.
h. Recognize the essential nature of pesticides in agricultural production and public health.

Unit 2   Pesticide Vocabulary

a. Differentiate between pesticide and other chemical
   groups ending in "-cide".
b. Explain the function of each major class in the pesticides vocabulary.
c. Select the precise class when referring to a particular
   pesticide.
d. Indicate the classes of pesticides used in greatest
   volume.
e. Explain the legal definitions of pesticides as they
   appear in federal and state laws.

Unit 3   Pesticide Chemistry

a. Define atom, molecule, and compound.
b. List the chemical elements and their symbols involved
   in pesticide chemistry.
c. Differentiate between organic and inorganic compounds.
d. Identify the three types of chemical formulas.
e. Recognize the common, proprietary and chemical names of
   pesticides.
f. Explain why common names are given to most pesticides.

Unit 4   Pesticide Formulations

a. Differentiate between technical grade and formulated
   pesticides.
b. Summarize the reasons for formulating pesticides.
c. Describe the common formulations of pesticides.
d. Differentiate between normal and invert emulsions.
e. List the advantages and disadvantages of formulations
   for specific uses.
f. Predict the relative rates of deposit on target from
   most common formulations.
g. Explain why most of the newer formulations are in the
   flowable or sprayable category.

LEARNING OBJECTIVES

Unit 5 Insecticides

a. Categorize the organochlorine, organophosphate and carbamate insecticides into their major sub-groupings.
b. Describe the mode of action for the organochlorine, organophosphate, carbamate, formamidine, synthetic pyrethroid, and remaining major classifications of insecticides.
c. Identify the classifications of insecticides based on chemical structures.
d. List the characteristics of persistent insecticides.
e. Describe the biochemical transmission of nerve impulses.
f. Explain synergism and antagonism as found in insecticidal actions.
g. List the botanical insecticides and describe their modes of action.
h. Describe the 4 generations of synthetic pyrethroids and some of their physical characteristics.
i. Explain the modes of action for the microbial insecticides.
j. Identify the 3 generations of insecticides and give examples.

Unit 6 Herbicides

a. List the general use patterns for herbicides other than for weed control in planted crops.
b. Outline the ways to achieve herbicidal selectivity with non-selective materials.
c. Classify herbicides as to contact or translocated effects.
d. Classify herbicides as to timing of applications.
e. Classify herbicides based on area covered during application.
f. Categorize the major classifications of organic herbicides based on chemical structures, and describe their modes of action.
g. Identify the most heavily used category of herbicides.

Unit 7 Fungicides and Bactericides

a. List the causal agents of plant diseases.
b. Explain why some plant diseases are difficult to control with chemicals.
c. Explain the delicate nature of controlling fungal diseases on plants.
d. Distinguish between the uses of protectants, therapeutants and eradicants.
e. Categorize the inorganic and major classifications of organic fungicides.
f. Describe the modes of action of the heavy metal and the more recent non-metallic fungicides.
g. Explain why resistance to the modern synthetic fungicides occurs readily, but not to the older heavy metal

fungicides.
h. Account for the sudden popularity of the new systemic
   fungicides.

Unit 8 Nematicides

a. Explain the need for highly-penetrating chemicals as
   nematicides.
b. List the ways nematodes can cause plant disease.
c. List the 4 categories of nematicides currently used in
   agriculture and their modes of action.
d. Account for the recent decline of the traditional
   volatile, halogenated nematicides.
e. Identify the categories containing the nematicides of
   the future.

Unit 9 Rodenticides

a. Describe the need for using rodenticides the world over.
b. List the 4 most common methods of controlling rodents.
c. Explain why poisoning is the most widely used method of
   rodent control.
d. Describe the modes of action for the commonly used
   rodenticides.
e. Explain why the early coumarin rodenticides are the
   safest to use, and the more recent coumarins may be the
   most hazardous.
f. Explain why sodium fluoroacetate can be handled only by
   specially trained persons.
g. List the primary reasons why a certain rodenticide may
   not be the material of choice.

Unit 10 Plant Growth Regulators

a. List the functions hormones play in plant growth and
   development.
b. Summarize the value of plant growth regulators to man.
c. List the synonyms of plant growth regulators.
d. List the 6 classes of plant growth regulators and
   general uses of each class.
e. Explain the modes of action for the more common growth
   regulators.

Unit 11 Defoliants

a. List the crops on which defoliants are used and explain
   the role of defoliants.
b. Define the conditions needed for successful defoliation.
c. Explain the modes of action for the commonly used
   defoliants.

Unit 12 Desiccants

a. Differentiate between the physical actions of defoliants
   and desiccants.
b. Explain the mode of action of desiccants.
c. Explain why defoliants are used in some instances and
   desiccants in others.

Unit 13 Pesticides and The Law

a. Explain the purposes of the Federal Insecticide Act of
   1910, FIFRA of 1947, the Miller Amendment of 1954, the
   Food Additives Amendment of 1958, the Delaney Clause of
   1958, and the FEPCA Amendment of FIFRA of 1972.
b. List the 8 basic provisions of FEPCA.
c. Define unclassified and restricted use pesticides.
d. Define certified applicator as used in FEPCA.
e. Outline the areas of proficiency required to be covered
   in the examination for certification of commercial applicators.
f. Identify the conditions permitting use of a pesticide
   inconsistent with its label.

Unit 14 Safe Handling and Use of Pesticides

a. List safety precautions for the employer, for selection
   of pesticides, for transporting, storing, handling and
   mixing, applying, and disposing of empty containers and
   unused pesticides.
b. Prepare a list of at least 10 safety rules applicable to
   his own work area and situation.
c. Identify the best sources of information in case of a
   pesticide emergency.

Unit 15 Pesticide Toxicity

a. Distinguish between toxicity and hazard as they relate
   to pesticide use.
b. List the two types of risks associated with the use of
   pesticides.
c. Define LD_{50} and name the units of weight used to express
   LD_{50}s.
d. Calculate the LD_{50} for an average-sized man if given the
   LD_{50} of a compound for a standard laboratory test animal
   having a similar physiology.
e. List the types of toxicity data available for pesticides
   and indicate which of these is most realistic or useful
   to the applicator or handler.

UNIT 1
PESTICIDES — CHEMICAL TOOLS FOR PEST CONTROL

The chemical tools used to control all kinds of pests are known as *pesticides*. To the farmer pests include insects and mites that feed on crops; weeds in fields and aquatic plants that clog irrigation and drainage ditches; plant diseases caused by fungi, bacteria and viruses; nematodes; snails and slugs; and rodents and birds that consume unbelievable quantities of plant seedlings and grain from fields and animal feedlots as well as from storage. To the urbanite pests include annoying and disease-carrying flies, mosquitoes, and cockroaches; moths that eat his woolens and beetles infesting his packaged foods; slugs, snails, aphids, mites, beetles, caterpillars and bugs feeding on his lawn, garden and ornamental plants; termites that feed on his dwellings; weeds in his lawn and garden; diseases marring and destroying his plants, algae clouding his swimming pool; mildew staining the shower curtain; rats and mice destroying goods and eating his food; and nuisance birds whose feces encrust sidewalks, window ledges and statues of his heroes.

In 1985 there were registered with the U. S. Environmental Protection Agency (EPA) about 960 pesticide active ingredients (ai), of which 240 were herbicides, 225 were insecticides, 170 were fungicides and nematicides, 35 were rodenticides, and 210 were disinfectants. These were sold in more than 25,000 products or formulations. In the United States alone 4.4 pounds of pesticides are used each year for every man, woman, and child to feed, clothe and protect them.

More than 50 basic manufacturers in the U. S. make pesticides of one kind or another. Of these, 30 manufacture almost all of the total U. S. production, with 14 firms accounting for about 85 percent of all pesticide sales. Most of the basic producers prepare one or more ready-to-use forms or formulations of their products. Another 3,300 formulators throughout the country prepare the 25,000 different products for retail sales.

Pesticides have become essential tools to the agriculturist and the urbanite alike. Just as the tractor, mechanical harvester, electric milker, and fertilizers are part of modern agricultural technology, so are pesticides. The urbanite, too, depends on pesticides, perhaps, more than he realizes: for control of algae in swimming pools,

weed control in his lawns, flea powder for pets, sprays
for controlling a host of garden and lawn insects and
disease, household sprays for ants and roaches, aerosols
for flies and mosquitoes, soil and wood treatment for termite protection, rodent bait for the occasional mouse or
rat, insecticide treatment of woolens at the dry cleaners
for clothes moth protection, and repellents to keep off
biting flies, chiggers, and mosquitoes while fishing and
camping.

The total of pesticides used in the United States in
1983 was estimated at 0.953 billion pounds of active ingredients valued at $6.05 billion at the user or retail level.
Of this, the agricultural market consumed 77%, industry and
government utilized 16%, while home and garden use amounted
to 7% of the volume in pounds.

Of the pesticides used in the United States, agriculture utilizes 77% of the herbicides, 73% of the insecticides,
71% of the fungicides, and 99% of the fumigants, rodenticides and molluscicides. In the distribution of these major
classes in agricultural usage, 61% is herbicides, 25%
insecticides, 6.5% fungicides and 7.5% others. (Aspelin &
Ballard, 1984). It is interesting to note that 94% of the
herbicides and 89% of the insecticides used in agriculture
are used on only four major crops: corn, cotton, soybeans,
and small grains, including sorghum (Panel A).

PANEL A

SHARE OF TOTAL AGRICULTURAL HERBICIDES AND INSECTICIDES
USED IN 1982 ON CORN, COTTON, SOYBEANS AND WHEAT

Crop	Share of pesticide used in agriculture (a.i.)	
	Herbicides	Insecticides
	---Percent---	
Corn	53.9	42.5
Cotton	3.8	23.9
Soybeans	27.7	15.4
Small grains (including sorghum)	8.6	7.2
Combined	94.0	89.0

(Delvo and Hanthorn, 1983)

Industrial and commercial use consists of pesticides used by pest control operators, turf and sod producers, floral and shrub nurseries, railroads, highways, utility right-of-ways, and industrial plant sites. Government use includes federal and state pest suppression and eradication programs, and municipal and state health protection efforts involving the control of disease vectors such as mosquitoes, flies, cockroaches, and rodents. Industry and government used 68 percent herbicides, 26 percent insecticides, 6 percent fungicides, and less than 1 percent others.

Recent studies show that 91 percent of households in the United States use some form of pesticides. These included use of retail products, professional pest control services, and pet preparations. In the home and garden the volume of pesticide usage by classes is 38 percent herbicides, 46 percent insecticides, 15 percent fungicides and less than 1 percent others. (Aspelin & Ballard, 1984)

The price paid by the homeowner for pesticides is naturally much higher than agriculture or industry. This is due mainly to the specialized formulations used and the small package purchases. In 1983 the average price per pound paid for pesticides was $7.68 for home and garden, $6.44 by industry and government, and $6.20 by agriculture.

World history contains innumerable examples of the mass destruction of crops by diseases and insects. In 1845-1851 the potato famine of Ireland occurred as a result of a massive infection of potatoes by a fungus, _Phytophthora infestans,_ commonly known as late blight. (There are several fungicides which would now easily control that disease with two or three properly timed applications.) This resulted in the loss of about a million lives and the cultural invasion of America by the Irish. Surprisingly, the infected potatoes were edible and nutritious, but a superstitious population refused to use diseased tubers. In 1930, 30% of the U. S. wheat crop was lost to stem rust, the same disease that destroyed three million tons of wheat in Western Canada in 1954.

The Panama Canal was abandoned in the 19th century by the French because more than 30,000 of their laborers died from yellow fever. Since the first recorded epidemic of the Black Death, Black Plague, Plague, Pale Horseman, or Bubonic Plague, it is estimated that more than 65 million persons have died from this disease transmitted by the rat flea. The number of deaths resulting from all wars appears paltry beside the toll taken by insect-borne diseases.

Until as recently as 1955, malaria infected more than 200 million persons throughout the world. The annual death rate from this debilitating disease has been reduced from 6 million in 1939 to less than a million today. Similar progress has been made in controlling other important tropical diseases such as yellow fever, African sleeping sickness, and Chagas' disease through the use of insecticides.

There is currently the ever-lurking danger to man from such diseases as encephalitis, typhus, relapsing fever, sleeping sickness, elephantiasis, and many others, which are transmitted by insects or mites.

Of equal importance are the agricultural losses from weeds. They deprive crop plants of moisture and nutritive substances in the soil. They shade the crop plants and hinder their normal growth. They contaminate harvested grain with seeds that may be poisonous for man and animals. In some instances, complete loss of the crop results from disastrous competitive effects of weeds.

In a 10 year study at the University of Illinois, herbicide treatments increased yields 20 percent on the average for corn and soybeans. Wheat yields were not significantly increased. The economics of these results showed a rate of return on herbicide expenditures of about $4.00 per $1.00 spent.

If it weren't for herbicides, we would still have 10 to 12 percent of our population working on farms, instead of the present 2 percent. Today's farms in the United States would quickly become perpetuating weed fields, consuming tremendous amounts of our human energy as so much of the rest of the world's population is employed. It has even today been estimated that more energy is expended for the weeding of crops than for any other single human task!

Pesticides are used by man as intentional additions to his environment in order to improve environmental quality for himself, his animals, and his plants. Pesticides are used in agriculture to increase the ratio of cost/benefit in favor of the grower and ultimately the consumer of food and fiber products--the public. Pesticides have contributed significantly to the increased productive capacity of the U. S. farmer, each of whom produced food and fiber for 4 persons in 1850, 25 in 1960, 38 in 1970, 60 persons in 1981 and 82 in 1985.

World population was estimated at 3.6 billion in 1970, 4.4 billion in 1980, and is expected to reach 5.4 billion in 1990, and 6.4 billion by 2000. These numbers are not intended to evoke gloom about our ability to support such a population but to suggest that there will be tremendous pressure to increase agricultural production since this extra population will have to be fed and clothed.

Even our current world food supply is inadequate. As much as 56% of the world's population is undernourished. And the situation is worse in third world countries, such as Ethiopia, Sudan, Chad, Mali and Niger, where an estimated 79% of the inhabitants are undernourished or starving.

Plants are the world's main source of food. They compete with about 80,000-100,000 plant diseases caused by viruses, bacteria, mycoplasma-like organisms, rickettsias, fungi, algae, and parasitic higher plants; 30,000 species of weeds the world over, with approximately 1,800 species causing serious economic losses; 3,000 species of nematodes that attack crop plants with more than 1,000 that cause damage; and over 800,000 species of insects of which 10,000 species add to the devastating loss of crops throughout the world.

Pesticides have rapidly evolved as extremely important aids to world agricultural production. It is estimated that insects, weeds, plant diseases and nematodes account for losses exceeding $20 billiom annually in the U. S. alone. The use of these pesticidal chemicals in agriculture makes it possible to save approximately an overall one-third of our crops. It is because of the economic implications of such losses and savings that pesticides have assumed their importance.

Up to one-third of the world's food crops are destroyed by pests during growth, harvesting and storage. Losses are even higher in emerging countries: Latin America loses to pests approximately 40% of everything produced. Cocoa production in Ghana, the largest exporter in the world, has been trebled by the use of insecticides to control just one insect species. Pakistan sugar production was increased 33% through the use of insecticides. The FAO has estimated that 50% of cotton production in developing countries would be destroyed without the use of insecticides.

Several good examples of specific increases in yields resulting from use of insecticides were determined in 1976-78, by controlling insects in test plots with insecticides and comparing the yields with adjacent plots in which the insects were allowed to feed and multiply uncontrolled. From a single insect pest species, under severe conditions, each of the major crops suffered substantial losses: Corn from the corn borer, 24 percent; soybeans from the Mexican bean beetle, 26 percent; wheat from the wheat mite, 79 percent; cotton from the bollworm complex, 79 percent; and potatoes from the European corn borer, 53 percent.

World losses from pests, diseases, and weeds are estimated at more than $100 billion annually. From all of this it becomes obvious to the reader that control of various harmful organisms has great importance for agriculture, industry, and public health. Thus pesticides become indispensable in feeding, clothing and protecting the world's population, which is predicted to reach nearly 7.0 billion by New Year's Day, A.D. 2000.

World losses from pests, diseases, and weeds are estimated at more than $100 billion annually. From all of this it becomes obvious to the reader that control of various harmful organisms has great importance for agriculture, industry, and public health. Thus pesticides become indispensable in feeding, clothing and protecting the world's population, which is predicted to reach nearly 7.0 billion by New Year's Day, A.D. 2000.

1.1   Name 8 pests of agriculture and 5 of suburbia.

      Agriculture:_____, _____, _____,

      _____, _____, _____, _____, _____.

      Suburbia: _____, _____, _____, _____, _____.
      -
      Agriculture: weeds; aquatic plants; plant diseases
                     caused by fungi, bacteria, viruses;
                     nematodes; snails; slugs; rodents;
                     birds.
      Suburbia: flies, mosquitoes, cockroaches, moths,
                beetles, termites, weevils, plant diseases,
                rats and certain birds.

1.2   Today there are   active ingredients registered
      with the Environmental Protection Agency. Of these
            are herbicides, _____ insecticides, and _____
      fungicides.
      - - - - - - - - - - - - - - - -
            herbicides   insecticides   fungicides

1.3   List four diseases of man or plants that changed his-
      tory. _____, _____,

      _____, and _____.
      -
      potato late blight, yellow fever, malaria, plague

1.4   What is the annual usage, in pounds, of pesticides
      in the U.S. today for each citizen? _____ pounds
      -
      4.4 pounds

1.5   Why can't we return to the agricultural practices of
      the "good old days"?
      _____
      _____
      -
      Lack of manpower on farms to do the work that machin-
      ery and pesticides have replaced.

1.6 How great a catastrophe would a severe outbreak of potato blight in Ireland be today?

_____

- -

No problem, since it could be easily controlled by the application of modern fungicides.

1.7 In your own words, what would be the result of withholding all pesticide use in the U.S. for just one year?

_____

- -

Generally, our overall food and fiber production would decline between 25 and 33%. Urban pests of all types could also become rampant, with potential outbreaks of several insect-borne diseases of man and his domestic animals.

1.8 Agriculture uses _____%, industry and government use _____%, while home and garden account for only _____% of the total pesticides used in the United States.

- -

77% agriculture   16% industry/govt.   7% home/garden

1.9 Not only have we become totally dependent on pesticides, but they are now essential in agricultural production and public health. Why?

_____

- -

Essential to feed, clothe, and protect from insect-vectored disease the 6.4 billion world population in the year, 2000.

UNIT 2
PESTICIDE VOCABULARY

With the importance of pesticides established and the need for knowledge of these essential but somewhat mysterious chemicals is recognized by the reader, let's begin the task of learning what pesticides are all about. First, the language of pesticides needs to be accurate. To one person a pesticide means the insecticide malathion, to another the herbicide atrazine, and to still another, the fungicide maneb. Each is correct, but only partly so for the uses of these three compounds are unrelated.

The word *pesticides* is an "umbrella word", in that it is a very broad term which covers a large number of more accurate names. Examine the umbrella used for ullustration. Notice that the entire canopy is labeled PESTICIDES, while the panels are each labeled differently, (INSECTICIDES, HERBICIDES, FUNGICIDES, etc.) These various generic words ending in *-cide* are included in the umbrella PESTICIDES, an all-inclusive term meaning literally *killers of pests*.

From the legal viewpoint pesticides are classed as "economic poisons" in most state and federal laws and are defined as "any substance used for controlling, preventing, destroying, repelling, or mitigating any pest."

Included in pesticides are groups of chemicals that do
not actually kill pests. However, because they fit rather
practically as well as legally into this umbrella word,
pesticides, they are included. Among these are chemical
compounds that stimulate or retard growth of plants and
sometimes insects, *growth regulators*; those that remove
leaves, *defoliants*; or speed the drying of plants, *desiccants*, and are used for mechanizing the work in harvesting
cotton, soybeans, potatoes and other crops.

The term pesticides also applies to compounds used as
repellents, attractants, and *insect sterilants* or *chemo-
sterilants*. However, these last groups do not precisely
fit the original definition but rather the legal definition
of pesticides.

In Panel B is a list of the most common pesticide
classes, their uses and derivations. Occasionally a word
will be generated to describe the use of a chemical for
which none of those listed in the table applies, for instance, biocide. This is a nondescript word usually found
in sensational journalism. It is commonly associated with
adverse effects of pesticides and often times used as a
synonym for pesticide. Biocide is not an accepted word in
the scientific vocabulary. It has no exact meaning, thus
is usually used to describe a chemical substance that kills
living things or material. Regardless of its presumed
meaning, it appears in sensational and emotional rather
than scientific writings.

Of all the groups listed in Panel B, those most widely
used in the greatest volume in producing our food and fiber
crops are the herbicides, insecticides, and fungicides, in
that order, as you may have guessed after reading Unit 1.

This is the end of the vocabulary section, which includes all of the commonly used pesticides, including
groups of chemicals that do not actually kill pests. But
because they fit practically as well as legally into this
umbrella word, pesticides, they are included.

Your vocabulary should now be considerably larger and
more precise than before. You can distinguish between pesticide and herbicide, fungicide and defoliant, or acaricide
and growth regulator. To be certain, briefly review the
table again before advancing to the questions.

PANEL B
PESTICIDE CLASSES,
THEIR USE AND DERIVATION

Pesticide Class	Function	Root-word Derivation
Acaricide	kills mites	Gr. $\frac{a}{}$ *akari*, mite or tick
Algicide	kills algae	L. $\frac{b}{}$ *alga*, seaweed
Avicide	kills or repels birds	L. *avis*, bird
Bactericide	kills bacteria	L. *bacterium* (Gr. bactron), a staff
Fungicide	kills fungi	L. *fungus*, (Gr. spongos) mushroom
Herbicide	kills weeds	L. *herba*, an annual plant
Insecticide	kills insects	L. *insectum*, insecre, cut or divided into segments
Larvicide	kills larvae (usually mosquito)	L. *lar*, mask or evil spirit
Miticide	kills mites	Synonymous with *Acaricide*
Molluscicide	kills snails and slugs (may include oysters, clams, mussels)	L. *molluscus*, soft- or thin-shelled
Nematicide	kills nematodes	L. *nematoda* (Gr. *nema*), thread
Ovicide	destroys eggs	L. *ovum*, egg
Pediculicide	kills lice (head, body, crab)	L. *pedia*, louse
Piscicide	kills fish	L. *piscis*, a fish
Predicide	kills predators (coyotes, usually)	L. *praeda*, prey
Rodenticide	kills rodents	L. *rodere*, to gnaw
Silvicide	kills trees and brush	L. *silva*, forest
Slimicide	kills slimes	Anglo-Saxon *slim*
Termiticide	kills termites	L. *termes*, wood-boring worm

PANEL B (Continued)

CHEMICALS CLASSED AS PESTICIDES
NOT BEARING THE — CIDE SUFFIX

Pesticide Class	Function
Attractants	attract insects
Chemosterilants	sterilize insects or pest vertebrates (birds, rodents)
Defoliants	remove leaves
Desiccants	speed drying of plants
Disinfectants	destroy or inactivate harmful micro-organisms
Feeding stimulants	cause insects to feed more vigorously
Growth regulators	stimulate or retard growth of plants or insects
Pheromones	attract insects or vertebrates
Repellents	repel insects, mites and ticks, or pest vertebrates (dogs, rabbits, deer, birds)

---

[a] Greek origin

[b] Latin origin

2.1 The word-ending or suffix -cide, means _____,
     thus pesticides are used to _____pests.
     ----

     kill or killer   kill

2.2 From this logic we conclude that insecticides are used
     to kill _____, rodenticides to kill _____,
     and herbicides to kill _____.
     ----

     insects   rodents   weeds

2.3 Acaricides (miticides and tickicides) are used
     against _____ and _____, algicides
     against _____ and _____, and fungicides
     against _____.
     ----

     mites, ticks   algae, aquatic weeds   disease fungi

2.4 Pesticides used to control the growth of plants are
     _____, while those used to
     defoliate plants are _____.
     ----

     growth regulators   defoliants

2.5 Materials used for drying green plants are known as
     _____, while those used to repel insects and
     mites are _____.
     ----

     desiccants   repellents

2.6 By volume the groups used in greatest quantity are
     _____, _____ and _____.
     ----

     herbicides, insecticides, and fungicides

2.7 In most state and federal laws pesticides are classed
     as _____ _____.
     ----

     economic poisons

UNIT 3
PESTICIDE CHEMISTRY

For a better understanding of some of the chemistry of pesticides it will be necessary to learn to identify the different elements with which pesticides are made. You may have had a course in chemistry, and if so, skip to Unit 4.

The earth and everything about us is composed of chemical *elements*. Oxygen, nitrogen, iron, sulfur and hydrogen are elements. Water, salt and malathion are *compounds*. The smallest part or unit of an element is the *atom*. An element is pure in the sense that all of its atoms are alike. It is not possible to change an element chemically.

3.1 Everything about us is made of _____.

----

    elements

3.2 The smallest unit of an element is the _____.

----

    atom

3.3 Examples of elements are _____,
_____, and _____.

----

    oxygen, nitrogen, iron, etc.

A *compound* is a combination of two or more different elements. When the different atoms combine or join they form a *molecule*, which is the smallest part or unit of a compound. Because a compound is a combination of different elements, chemical changes can be made in it by changing the combinations of elements.

3.4   The combination of two or more elements is a
      _____.

- -

      compound

3.5   The smallest unit of a compound is the _____.

- -

      molecule

      Elements have a certain way of combining with other
elements, and only certain elements will combine.
      Chemical compounds are combinations of two or more
elements bound together by chemical bonds. The most familiar compound would be water, a mixture of hydrogen and oxygen, bound together by chemical bonds, which we recognize
as H_2O. To use the proper chemical terminology, H_2O is two
atoms of hydrogen bonded to one atom of oxygen to form one
molecule of water.

3.6   The smallest unit of water that can exist is
      _____.

- -

      one molecule

3.7   Pesticides are compounds, so they too must be made of
      different elements or atoms bonded together chemical-
      ly. From the above description we then know that the
      smallest unit of any pesticide that can exist is a
      _____.

- -

      molecule

      Chemists have found it helpful to use a sort of chem-
ical shorthand in writing of these elements, and a *symbol*
was chosen for each. This makes both the writing and the
setting of type for printed work simpler and easier to
read. This symbol is often the first letter or the first
plus another letter in the element name, e.g. hydrogen =
H, chlorine = Cl.

3.8 A shorthand method of writing the chemical name is
     the _____.

- -

     symbol

     On the next page in Panel C are shown some of the
elements or atomic names and the chemical symbols used
when writing or illustrating molecules. The Panel is
incomplete since it contains only those elements illustrated or referred to in the text following. Surprisingly
few additional elements are found in the large array of
pesticides available.
     As you study the pesticides, you will notice that only
21 of the more than 105 elements are used in their con-
struction. We can further reduce the list by considering
those used most frequently in pesticides: carbon, hydrogen, oxygen, nitrogen, phosphorus, chlorine, and sulfur.
Some, however, may include the metallic and semi-metallic
elements, iron, copper, mercury, zinc, arsenic, and others.
And because only about one-half of the elements shown in
Panel C make up the bulk of the most-used pesticides it is
a simple matter to learn their symbols.

3.9 Write the symbols for:
     Oxygen _____   Carbon _____   Hydrogen _____   Nitrogen_____
     Chlorine_____ Sulfur _____   Phosphorus_____ Iron _____
     Copper _____   Bromine_____   Zinc _____

- -

     Oxygen   O   Carbon   C   Hydrogen   H   Nitrogen N
     Chlorine Cl Sulfur   S   Phosphorus P   Iron   Fe
     Copper   Cu Bromine Br Zinc   Zn

     Some 95% of the pesticides studied in this book are
made from these 11 elements. If you missed more than one
symbol, review the last three pages and complete the blanks
again.
     Essentially all of the pesticides are *organic* com-
pounds, that is, they contain *carbon* in their molecules.
A very few contain no carbon as you will see, and are
termed *inorganic* compounds.

3.10 Organic compounds contain _____.

- -

     carbon

PANEL C
CHEMICAL ELEMENTS
FROM WHICH PESTICIDES ARE MADE[a]
(This material should be used for questions 3.11 — 3.14)

Atomic or Element Name	Symbol	Symbol Derivation[b]	Pesticide Example[c]
Arsenic	As		cacodylic acid (H)
Boron	B		sodium borate (H)
Bromine	Br		methyl bromide (Fum)
Cadmium	Cd		cadmium chloride (F)
Carbon	C		any organic pesticide
Chlorine	Cl		chlordane (I)
Copper	Cu	cuprum	Bordeaux mixture (F)
Fluorine	F		sufuryl fluoride (Fum)
Hydrogen	H		any organic pesticide
Iron	Fe	ferrum	ferbam (F)
Lead	Pb		lead arsenate (I)[d]
Magnesium	Mg		magnesium fluosilicate (I)
Manganese	Mn		maneb (F)
Mercury	Hg	hydrargyrum	mercuric chloride (F)
Nitrogen	N		methomyl (I)
Oxygen	O		organophosphate insecticides
Phorphorous	P		organophosphate insecticides
Sodium	Na	natrium	sodium chlorate (D)
Sulfur	S		tetradifon (M)
Tin	SN	stannum	cyhexatin (M)
Zinc	Zn		zineb (F)

[a]This list is incomplete and includes only those elements which will be used subsequently.

[b]Some atomic symbols are based on the Latin names.

[c]Letter in brackets following pesticide denotes pesticide class, e.g. H = herbicide, I = insecticide, F = fungicide, D = defoliant/desiccant, M = miticide, Fum = fumigant.

[d]Example only; pesticides containing lead are no longer registered for use in the U. S.

3.11 Inorganic compounds do not contain _____ .

- -

      carbon

_____

3.12 One of the simplest organic compounds would be carbon
      dioxide, CO_2, because it contains _____ .

- -

      carbon

_____

3.13 List three simple inorganic compounds, using the
      elements shown in Panel C.

      (1)_____ (2)_____ , (3)_____

- -

      (1) H_2O (2) SO_2 (3) NaCl and H_2S, HCl, $ZnCl_2$

_____

3.14 In summary, pesticides are usually _____
      compounds containing two or more _____ in
      their molecules.

- -

      organic   elements

_____

3-A CHEMICAL FORMULAS

      There are several kinds of chemical formulas that
should be distinguished from each other since all are used
in this manual.

      The *molecular formula* uses the symbols of the elements
to indicate the number and kind of atoms in a molecule of
the compound, as H_2O for water and C_6H_6 for benzene. The
structural formula is written out, using symbols to indicate the way in which the atoms are located relative to
one another in the molecule.

      Structural formulas for water and benzene would be
presented in this way:

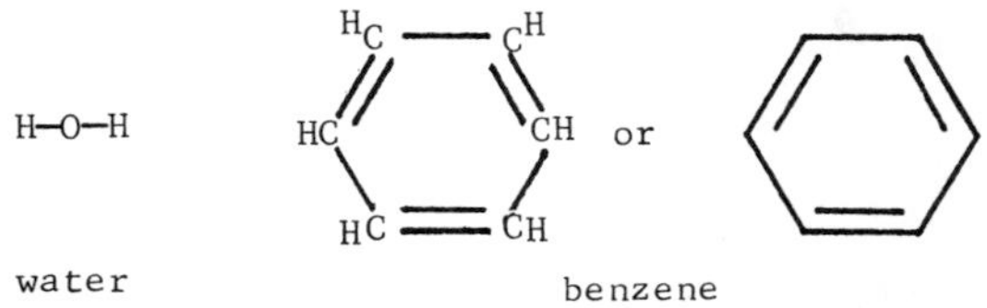

              water   benzene

Let us examine a bit further this 6-membered ring-structure, benzene. As presented with nothing attached, it is benzene, and for ease of representation, it is usually indicated as a hexagon with double bonds. The hydrogens are usually not printed as shown. When other groups have replaced one or more of the hydrogens, the ring is referred to as the phenyl radical rather than the benzene radical. To illustrate, turn to page 41 where the structure of DDT is shown. Note that the chemical name of DDT contains the words "di-phenyl", meaning two benzene rings with other groups attached.

3.15   When benzene becomes a part of a larger molecule, the benzene ring will be referred to as _____.

- -

phenyl

Because the benzene ring or phenyl radical is found frequently in organic chemicals, and because it is an awkward structure for printers to set from type, simpler ways of representation have been devised, and are commonly found in pesticide literature. One of these has already been shown. The four remaining designs are (1) the hexagon containing a circle or oval, (2) the hexagon containing a broken circle, (3) the hexagon with the three double bonds extending to its sides, and finally, (4) the Greek letter Phi, φ or Φ.

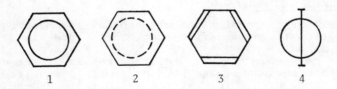

1   2   3   4

3.16 Illustrate the 5 ways that the phenyl radical appears in pesticide literature.

- -

3.17 In this frame designate the molecular formula for
carbon dioxide_____; the structural formula for
carbon dioxide_____; the molecular formula for arsenic acid_____; and the structural formula for arsenic acid_____.

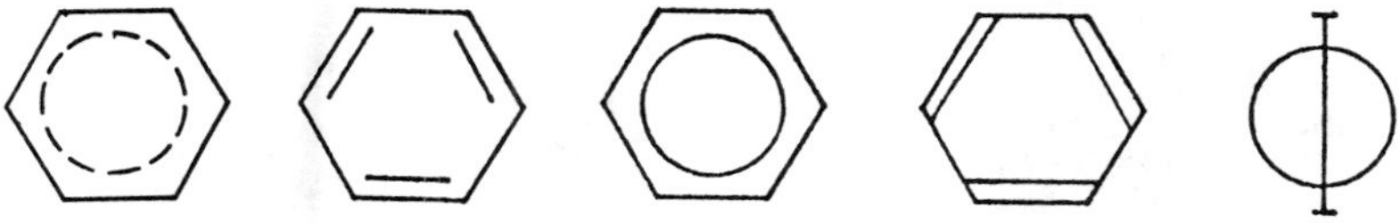

A. $O=C=O$ B. CO_2 C. H_3AsO_4 D. $HO-\overset{\overset{O}{\|}}{\underset{\underset{H}{O}}{As}}-OH$

- -

molecular carbon dioxide   B
structural carbon dioxide   A
molecular arsenic acid   C
structural arsenic acid   D

Structural formulas do not show the actual shape of
organic molecules, but it is not essential that the shape
and spatial design be known to become familiar with these
structures. The structural formula will be shown for most
of the compounds discussed in this manual.
 Three-dimensional or stereo-formulas are written so
that the reader can see the depth and spatial conformation
of molecular structures. Writing the structural formula
on paper does not make the molecule flat, since formulas
are two-dimensional representations of three-dimensional
objects.

3.18 In reality, would a molecule exist as a two- or
three-dimensional object?

- -

three-dimensional

Examples of two structural formulas converted to stereo formulas:

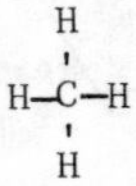

methane    stereo methane

norbornene 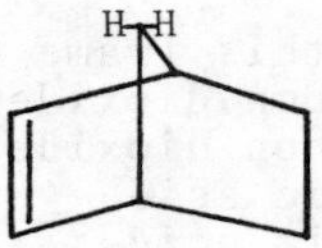   stereo norbornene

---

3.19 The three types of formulas used in illustrating compounds are _____, _____, and _____.

- -

molecular, structural, and 3-D or stereo

3-B   NOMENCLATURE

Knowledge of pesticides involves, among other things, learning about their structures and their names or nomenclature. For example, let's use Trithion[R] for illustration:

(I)   (II)

CARBOPHENOTHION   (Trithion®)

(III) 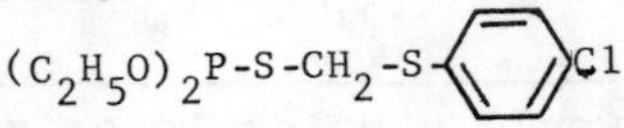 $(C_2H_5O)_2P\text{-}S\text{-}CH_2\text{-}S$⟨　　⟩Cl

(IV) S-[(p-chlorophenylthio) methyl]O,O-diethyl phosphorodithioate

(V) $C_{11}H_{16}ClO_2PS_3$

The name at the top, carbophenothion (I) is the common name for the compound. Common names are selected officially by the appropriate professional scientific society and approved by the American National Standards

Institute (formerly United States of America Standards
Institute) and the International Organization for Stan-
dardization. Common names of insecticides are selected by
the Entomological Society of America; herbicides by the
Weed Science Society of America; and fungicides by the
American Phytopathological Society. The proprietary name
(II), trade name or brand name, for the pesticide is given
by the manufacturer or by the formulator. It is not un-
common to find six or more brand or trademark names given
to a particular pesticide by various formulators. To
illustrate, Trithion[R] is also known as Garrathion[R],
Dagadip[R], and R-1303. The latter is likely to be a code
number assigned to the compound by the basic manufacturer
when it was first synthesized in the laboratory, but not
always. Common names are assigned to avoid the confusion
resulting from the use of several trade names, as just
illustrated. The structural formula (III), as mentioned
earlier, is the printed picture of the pesticide molecule.
The long chemical name (IV) beneath the structural formula
is just that, the scientific or chemical name. It is
usually presented according to the principles of nomenclature used in Chemical Abstracts, a scientific abstract-
ing journal which is generally accepted as the world standard for chemical names. Surprisingly, these sometimes
change in time, following new rules of naming chemicals.
And finally, (V) when used, is the molecular or *empirical*
formula which indicates the various numbers of atoms for
comparative purposes. Beyond this, we will use the molecular formula only when the structural formula is not known.

___

3.20 Why are common names assigned to most pesticides?

- -

    To avoid confusion caused by using its numerous trade
    names.

___

3.21 Of what value is the chemical name of a pesticide?

- -

    It clearly distinguishes one pesticide from another,
    is understood in all languages by chemists, and
    enables us to draw the structure by following the
    rules of chemical naming.

___

3.22 In this herbicide example, assign the appropriate titles to the blanks:

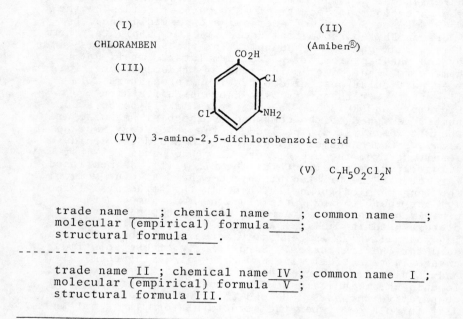

(I)

CHLORAMBEN

(II)

(Amiben®)

(III)

CO_2H

Cl

Cl

NH_2

(IV) 3-amino-2,5-dichlorobenzoic acid

(V) $C_7H_5O_2Cl_2N$

trade name____; chemical name____; common name____; molecular (empirical) formula____; structural formula____.

- -

trade name _II_ ; chemical name _IV_ ; common name _I_ ; molecular (empirical) formula _V_ ; structural formula _III_.

UNIT 4
PESTICIDE FORMULATIONS

Pesticides must be formulated before they are ready to be used.

Once the pesticide is manufactured in its relatively pure form, the *technical grade* material, whether herbicide, insecticide, fungicide or other classification, it is then formulated. That is, it is processed into a usable form for direct application, or for dilution followed by application. The *formulation* is the form in which the pesticide is sold for use. The technical grade material may be formulated by the *basic manufacturer* or sold to a formulator. The formulated material will be sold under the formulator's brand name or it may be custom-formulated for another firm.

4.1 The basic manufacturer produces the _____ _____ material.

- -

     technical grade

4.2 A pesticide is _____ into a usable form after manufacture.

- -

     formulated

4.3 The usable form of pesticide is a _____.

- -

     formulation

Formulation is the processing of a pesticidal compound by any method which will improve its properties of *storage, handling, application, effectiveness,* or *safety*. The term formulation is usually reserved for commercial preparation prior to actual use, and does not include the final dilution in application equipment.

4.4 Formulation of a pesticide improves its _____,
     _____, _____, _____, or _____.
----

     storage, handling, application, effectiveness, safety

     For final acceptance by the grower, commercial appli-
cator, or home owner, a pesticide must be effective, safe
and easy to apply, and generally economical. Pesticides
then, are formulated into many usable forms for satisfactory storage, for effective application, for safety to
the applicator, bystander or neighbor, and the environment,
for ease of application with readily available equipment,
and for economy. This is not always simply accomplished,
due to the chemical and physical characteristics of the
technical grade pesticide. For example: some materials in
their "raw" or technical condition are liquids, others are
solids; some are stable to air and sunlight, while others
are not; some are volatile, others not; some are water soluble, some oil soluble, and others may not be soluble in
either water or oil. These chacteristics pose problems to
the formulator, since the final formulated product must
meet the specifications of federal and state regulations
and must meet the standards of acceptability by the user.

4.5 In your own words explain the need for various formu-
     lations of pesticides.

     _____
     _____
     _____

     Stability and physical condition of technical material
     and its physico-chemical properties.

     Fully 95% of all pesticides used in the United States
in 1986 have been manufactured in the formulations appearing in the simplified classification presented in Panel D.
Familiarity with the more important formulations is essential to anyone working with pesticides. We will now examine the major formulations used in agriculture, structural
pest control, and by the homeowner.

4-A SPRAYS

     Emulsifiable Concentrates (EC). Formulation trends
shift with time and need. Traditionally, pesticides have
been applied as water sprays, water suspensions, oil
sprays, dusts, and granules. Spray formulations are prepared for insecticides, herbicides, miticides, fungicides,
algicides, growth regulators, defoliants, and desiccants.

PANEL D
COMMON FORMULATIONS OF PESTICIDES[a]

---

1. Sprays (insecticides, herbicides, fungicides)

   a. Emulsifiable concentrates (also emulsible concentrates)
   b. Water-miscible liquids, sometimes referred to as liquids
   c. Wettable powders
   d. Water-soluble powders, e.g., prepackaged, tank drop-ins, for agricultural and pest control operator use
   e. Oil solutions, e.g., barn and corral ready-to-use sprays, and mosquito larvicides
   f. Soluble pellets for water-hose attachments
   g. Flowable or sprayable suspensions
   h. Flowable microencapsulated suspensions, e.g., Penncap M[R]
   i. Ultralow-volume concentrates (agricultural and forestry use only)
   j. Fogging concentrates, e.g., public health mosquito and fly abatement foggers

2. Dusts (insecticides, fungicides)

   a. Undiluted toxic agent
   b. Toxic agent with active diluent, e.g., sulfur, diatomaceous earth
   c. Toxic agents with inert diluent, e.g., home garden insecticide-fungicide combination in pyrophyllite carrier
   d. Aerosol dust, e.g., silica aerogel in aerosol form

3. Aerosols (insecticides, repellents, disinfectants)

   a. Pushbutton
   b. Total release

4. Granulars (insecticides, herbicides, algicides)

   a. Inert carrier inpregnated with pesticide
   b. Soluble granules, e.g., dry flowable herbicides

5. Fumigants (insecticides, nematicides, herbicides)

   a. Stored products and space treatment, e.g., liquids, gases, moth crystals
   b. Soil treatment liquids that vaporize

6. Impregnates (insecticides, fungicides, herbicides)

   a. Polymeric materials containing a volatile insecticide, e.g., No-Pest Strips[R], pet collars
   b. Polymeric materials containing non-volatile insecticides, e.g., pet collars, adhesive tapes, pet tags, livestock eartags

PANEL D (Continued)

---

    c. Shelf papers containing a contact insecticide
    d. Mothproofing agents for woolens
    e. Wood preservatives
    f. Wax bars (herbicides)
    g. Insecticide soaps for pets

7. Fertilizer combinations with herbicides, insecticides, or fungicides

8. Baits (insecticides, molluscicides, rodenticides, and avicides)

9. Slow-release insecticides (see Panel E)

    a. Microencapsulated materials for agriculture, mosquito abatement, and household
    b. Paint-on lacquers for pest control operators and home-owners
    c. Interior latex house paints for home use
    d. Adhesive tapes for pest control operators and home-owners
    e. Resin strips containing volatile organophosphate fumigant, e.g., No-Pest Strips[R]

10. Insect repellents

    a. Aerosols
    b. Rub-ons (liquids, lotions, paper wipes, and sticks)

11. Insect attractants

    a. Food, e.g., Japanese beetle traps, ant and grasshopper baits
    b. Sex lures, e.g., pheromones for agricultural and forest pests

12. Animal systemics (insecticides, parasiticides)

    a. Oral (premeasured capsules or liquids)
    b. Dermal (pour-on or sprays)
    c. Feed-additive, e.g., impregnated salt block and feed concentrates

---

[a]This list is incomplete, but does contain most of the common formulations.

Consequently, 75% of all pesticides are applied as sprays.
The bulk of these are currently applied as water emulsions
made from emulsifiable concentrates.

4.6 Most pesticides are applied as _____ and
     the bulk of these are made from which formulation?

- -

     sprays   emulsifiable concentrates

    Emulsifiable concentrates, or emulsible concentrates,
are concentrated oil solutions of the technical grade
material with enough emulsifier added to make the concentrate mix readily with water for spraying. The emulsifier is a detergentlike material that makes possible the
suspension of microscopically small oil droplets in water
to form an emulsion.

4.7 Concentrated oil solutions of a pesticide and emulsi-
     fier are the _____ _____ formulation.

- -

     emulsifiable concentrate

4.8 The prime ingredient that makes possible the mixing of
     an emulsifiable concentrate and water is the

     _____.

- -

     emulsifier

    When an emulsifiable concentrate is added to water,
the emulsifier causes the oil to disperse immediately and
uniformly throughout the water, if agitated, giving the
water an opaque or milky appearance. This oil-in-water
suspension is a normal emulsion. There are a few rare
formulations of invert emulsions, which are water-in-oil
suspensions, and are opaque in the concentrated forms,
resembling salad dressing or face cream.

4.9 Oil-in-water suspensions are _____ emulsions.

- -

     normal

4.10 Water-in-oil suspensions are _____ emulsions.
----

        invert

        Emulsifiable concentrates, if properly formulated,
should remain suspended without further agitation for at
least one day after dilution with water. A pesticide concentrate that has been held over from last year can be
easily tested for its emulsible quality by adding one
ounce to a quart of water and allowing the mixture to stand
after shaking. The material should remain uniformly suspended for at least 24 hours with no precipitate. If a
precipitate does form the same condition will occur in the
spray tank, resulting in clogged nozzles and uneven application. Additional emulsifier can be obtained from the
formulator and added to the concentrate at the rate of
0.25 to 0.5 lbs per gallon of out-dated concentrate. If
the emulsion remains suspended for even a few hours, it
can be sprayed if agitated. (Note: there are three kinds of
emulsifiers, anionic, cationic and nonionic. It is essential to use the right one.)

4.11 A usable emulsifiable concentrate should remain sus-
     pended with no precipitate for at least _____.
----

        24 hours

        Each gallon of emulsifiable concentrate contains from
4 to 7 pounds of petroleum solvent. This is usually one
of the more expensive aromatic solvents such as xylene.
In the past six years petroleum solvents have increased
in price 300 percent or more, increasing significantly the
costs of formulation and consequently the cost of the formulated product to the buyer. These increases are not likely to stabilize within the foreseeable future, and formulators are searching diligently for other, more economical
formulations for their products.
        Water Miscible Liquids are exactly that, water-mixable
or water-miscible. The technical grade material may be
water-miscible initially or it may be alcohol-miscible.
These formulations resemble the emulsifiable concentrates
in viscosity and color, but do not become milky when
diluted with water. These may have several abbreviations:
S = solution, SC = soluble concentrate, L = liquid, and
WSC = water soluble concentrate.

4.12 How does a water-miscible liquid differ in appearance from an emulsifiable concentrate after dilution?

_____

_____

- -

Water-miscible remains clear after dilution.

_____

Wettable Powders (WP) are essentially dusts containing high concentrations of pesticide with a wetting agent to facilitate the mixing of the powder with water before spraying. The technical material is added to the inert diluent, in this case a finely ground talc or clay, in addition to a wetting agent, similar to a dry soap or detergent, and mixed thoroughly in a ball mill. Without the wetting agent, the powder would float when added to water and the two would be almost impossible to mix. Because wettable powders usually contain from 50% to 75% clay or talc, they sink rather quickly to the bottom of spray tanks unless the spray mix is agitated constantly.

_____

4.13 Wettable powders are mixtures of three ingredients, _____, _____, and _____.

- -

   dust   pesticide   wetting agent

_____

Many of the insecticides sold for garden use are in the form of wettable powders because there is very little chance that this formulation can burn foliage, even at high concentrations. This is not true for emulsifiable concentrates, since the original carrier is usually an aromatic solvent, which in relatively moderate concentrations can cause foliage burning at high temperatures ($>90°$F).

Flowable or Sprayable Suspensions (F or S) exemplify an ingenious solution to a formulation problem. Earlier it was stated that some pesticides are soluble in neither oil nor water. They are soluble in one of the exotic solvents, however, making the formulation quite expensive and perhaps pricing it out of the marketing competition. To handle the problem, the technical material is blended with one of the dust diluents and a small quantity of water in a mixing mill, leaving the pesticide-diluent mixture finely ground but wet. This "pudding" mixes well with water and can be sprayed with the same tank-settling characteristic as the wettable powders.

A second form of flowable is the blending of the finely ground insecticide carbaryl with molasses. This formulation (Sevimol[R]) reduces to some extent its drift off-target during aerial application, increases its adherence to

30   PESTICIDE FORMULATIONS

foliage, thus reducing removal by rain, and increases mor-
tality of moths that are attracted to feed on the molasses.
    A third flowable is made by mixing an emulsifiable
concentrate, containing a very high percentage of a water-
stable toxicant and a thickener, with 2 to 4 volumes of
water. This results in a thick, concentrated normal emulsion, which is then diluted for use with the appropriate
volume of water before using. This formulation resulted
from the need for an emulsifiable concentrate made with the
least solvent possible to avoid foliage burn (phytotoxicity)
in citrus groves. These formulations may occasionally be
identified as spray concentrates and are abbreviated SC.
    A fourth is the flowable microencapsulated (FM) form-
ulation. To produce these, the insecticide is incorporated
by a special process in small, permeable spheres of a poly-
mer or plastic, 15 to 20 μm (micrometers, 10^{-6}m) in diameter. These spheres are then mixed with wetting agents,
thickeners, and water to give the desired concentration of
insecticide in the flowable, usually 2 pounds per gallon.
These are discussed in more detail in the section on Slow-
Release Insecticides.
    Sevin^R XLR Plus, a fifth modification of flowables,
was introduced in 1985. It is an improved water- or rain-
resistant formulation containing latex as a binding and
sticking agent, which extends residual activity to 7-10
days. Additionally, its hazard to honeybees is reduced to
one-fourth that of the wettable powder because of its adherence to foliage rather than to foraging bees. Particle
size of XLR is reported to range from 5-10 μm diameter ver-
sus the 20 μm average particle size for the 80 percent wettable powder.
    A sixth modification is the dispersible grain or "dry
flowable". These are small granules which either dissolve
in water or immediately disperse. No shaking or stirring
is necessary before mixing; none remains in the container;
and there are no problems with overwinter storage in cold
climates, making shelf life almost unlimited. (Because of
their unique name, "dry flowables" have been included here
rather than in the Granular Pesticides category.)

4.14 Which of the formulations is undergoing the greatest
     changes? Why?

     _____ . _____

     _____ .

- -

     flowables   Petroleum costs are forcing formulations
                 to find more economical methods of packag-
                 ing their products.

Water Soluble Powders (SP) are properly titled and self explanatory. Here, the technical grade material is a finely ground water-soluble solid and contains nothing else to assist its solution in water. It is simply added to the spray tank where it dissolves immediately. Unlike the wettable powders and flowables, these formulations do not require constant agitation; they are true solutions and form no precipitate. Because of their sometimes dusty quality, soluble powders may be packaged in convenient, water-soluble bags to be dropped unopened into the spray tank.

4.15 Water soluble powders form _____ _____ in water and do not require _____ _____.

- -

     true solutions   constant agitation

Oil solutions, in their commonest form, are the ready-to-use-household and garden insecticide sprays sold in an array of bottles, cans, and plastic containers, all usually equipped with a handy spray atomizer. They also may be sold as oil concentrates of the pesticide to be diluted with oil before application or they may be sold in the dilute, ready-to-use form. In either case, the compound is dissolved in oil and is applied as an oil spray; it contains no dust diluent, emulsifier, or wetting agent. Oil solutions may be used as roadside weed sprays, for marshes and standing pools to control mosquito larvae, in fogging machines for mosquito and fly abatement programs, or for household insect sprays purchased in the supermarket. There is no abbreviation commonly used for oil solutions.

4.16 Oil sprays usually contain only the two ingredients, _____ and _____.

- -

     technical grade pesticide and oil.

Ultra-Low-Volume Concentrates (ULV) are usually the technical product in its original liquid form, or if solid, the original product dissolved in a minimum of solvent. They are usually applied without further dilution by special aerial or ground spray equipment that limits the volume from one-half pint to a maximum of one-half gallon per acre as an extremely fine spray. The ULV formulations are used where good results can be obtained while economizing through the elimination of the normally high spray volumes which vary from 3 to 10 gallons per acre. This

technique has proved extremely useful where insect control
is desired over vast areas. ULV sprays result in very
small droplets which may cause a drift problem. A recent
addition to ULV formulations has been the aerial application of pyrethroid insecticides at 0.02 to 0.2 pound active
ingredient (ai) in 1 quart of semi- or once-refined cottonseed or soybean oil per acre. The only crops where this
combination has been used at this writing are cotton and
soybeans. Malathion is registered in some states for use
on cotton when diluted with cottonseed oil.

4.17 ULV sprays are applied from a minimum of _____
     up to a maximum of _____ per acre.

     1/2 pint   1/2 gallon

4-B. DUSTS

     Historically dusts (D) have been the simplest formula-
tions of pesticides to make and the easiest to apply.
Examples of the *undiluted toxic agent* are sulfur dust used
in agriculture, and one of the household cockroach dusts,
boric acid. An example of the *toxic agent with active
diluent* would be one of the agricultural insecticides which
has sulfur dust as its *carrier* or diluent. A *toxic agent
with an inert diluent* is the most common type of dust formulation in use today, both in the home garden and in agriculture. Insecticides and fungicides are applied in this
manner, with the carrier being an inert clay such as pyrophyllite. The last type, the aerosol dust, is a finely
ground silica in a liquefied gas propellant that can be
directed into crevices of homes and commercial structures
for insect control.

4.18 The most commonly used dust formulation has an
     _____ diluent. A synonym for diluent
     is _____.
----

     inert   carrier

     Despite their ease in handling, formulation, and ap-
plication, dusts are the least effective (and ultimately
the least economical) of the pesticide formulations. The
reason being that dusts have a very low rate of deposit on
the target. In agriculture, for instance, an aerial application of a standard dust formulation of insecticide

will result in 10% to 40% of the material deposited on
the crop. The remainder drifts upward and down wind.
Psychologically, dusts are annoying to the non-grower who
sees these great billowing dust clouds resulting from an
aerial application, in contrast to the grower who believes
he is receiving thorough coverage for the very same reason.
Under similar circumstances, an aerial application of a
water emulsion spray will deposit 50% to 90% of the pesticide on target.

4.19 Why are sprays generally more effective than dusts?

Sprays leave a much higher rate of deposit on target
than dusts.

4-C GRANULAR PESTICIDES

Granular pesticides (G) overcome the disadvantages of
dusts in their handling characteristics. The granules are
small pellets formed from various inert clays and sprayed
with a solution of the toxicant to give the desired content. After the solvent has evaporated, the granules are
packaged for use. Granular materials range in size from
20 to 80 mesh, which refers to the number of grids per inch
of screen through which they will pass. Only insecticides,
and a few herbicides, are formulated as granules. They
range from 2% to 25% active ingredient, and are used mostly
in agriculture though some granular formulations of plant
systemic insecticides are made for use on ornamentals.
Granular treatments can be made at virtually any time of
day, since they can be applied aerially in winds up to 20
mph without problems of drift, an impossible task with
sprays or dusts. They also lend themselves to soil application in the drill at planting time to protect the roots
from insects or to introduce a systemic to the roots for
transport to above-ground parts. Microgranules, mesh size
ranging from 60 to 80 mesh, have been used on an experimental basis on cotton and field crops. They have the advantages of both dusts and granules.

4.20 In granular formulations mesh refers to

_____.

----

screen grid divisions/inch

4.21 Approximately how many 20-mesh granules would fit
     into a one-cubic-inch container? _____
     40-mesh? _____

     8000   64,000

4-D AEROSOLS

     We live in an aerosol culture; bug bombs, hair sprays,
underarm deodorants, home deodorizers, oven cleaners, mace,
shaving cream, lubricants, mouth fresheners, furniture
polish, car wax, electrostatic inhibitors, adhesives,
starch, fabric finishers, pruning sealer, spot removers,
engine cleaner, motor starting fluids, water repellents,
window sprays, repellents, paints, garbage can and shower-
tub disinfectants, and supremely, the foot and crotch
sprays of anti-itch remedies.
     The pushbutton variety of insecticide was developed
during World War II for the GIs. More recently the total-
release aerosol has been designed to discharge its entire
contents in a single application, and are available for
homeowners as well as for commercial operators. The nozzle
is depressed and locked into place, permitting the container's total emission while the occupants remain away for
a few hours. Of the 2.4 billion aerosol units manufactured
in 1984, 8% were insect sprays.
     They are effective only against flying and crawling
insects and provide little residual effect. How do aerosols
work? The active ingredients must be soluble in the volatile, petroleum solvent in its pressurized condition. The
pressure is provided more recently from carbon dioxide,
replacing the ecologically suspect fluorocarbon propellant.
When the petroleum solvent is atomized, it evaporates
rapidly, leaving the micro-droplets of toxicant suspended
in air.
     Aerosols commonly produce droplets well below 10 μm in
diameter, which are respirable. This means that they will
be absorbed by alveolar tissue in the lungs rather than
impinging in the bronchioles, as do larger droplets. Consequently, aerosols of all varieties should be breathed as
little as possible.

4.22 Why are aerosols not the preferred insecticide for
     crawling household insects?

     Provide little residual effect

4-E FUMIGANTS

    Soil fumigants are used in horticultural nurseries,
greenhouses, and on high-value cropland to control nematodes, insect larvae and adults, and sometimes to control
diseases and weed seeds. Depending on the fumigant, the
treated soils may require covering with plastic sheets for
several days to retain the volatile chemical, allowing it
to exert its maximum effect.
    Grain and other stored products fumigants are fre-
quently packaged in small, easy-to-use cans, that are opened
and dropped into grain bins. As with all fumigants, they
contain a warning agent to prevent accidental exposure to
dangerous fumigant vapor concentrations.

4-F IMPREGNATING MATERIALS

    Impregnating materials described in this book include
only treatment of woolens for mothproofing and timbers
against wood-destroying organisms. For several years,
woolens and occasionally leather garments have been moth-
proofed in the final stage of dry cleaning (usually chlorinated solvents). The last solvent rinse contains an ultra-
low concentration of the chlorinated, biodegradable insecticide, methoxychlor, which has good residual qualities
against moths and leather-eating beetle larvae. Railroad
ties, telephone and light poles, fence posts, and other
wooden objects that have close contact with or are actually
buried in the ground soon begin to deteriorate as a result
of attacks from fungal decay microorganisms and insects,
particularly termites, unless treated with fungicides and
insecticides. Such treatment permits poles to stand for 40
to 60 years that would otherwise have been replaced in 5 to
10 years. The insecticides of choice for wood exposed to
potential termite damage are dieldrin and chlordane.

4-G FERTILIZER COMBINATIONS

    Fertilizer combinations are formulations fairly famili-
iar to the urbanite who has purchased a lawn or turf fertilizer that contained herbicide for crabgrass control, insecticides for grubs and sod webworms, or a fungicide for
numerous lawn diseases. Fertilizer-insecticide mixtures
have been made available to growers, particularly in the
corn belt, by special order with the fertilizer distributor.
The fertilizer and insecticide can then be applied to the
soil during planting in a single, economical operation.

4-H SLOW-RELEASE PESTICIDES

    Slow-release pesticides are relatively new and few in
number. Insecticides are the only group to be so formulated. The first to appear was the Shell No-Pest Strips[R]
in 1963, which contained dichlorvos. This volatile organo-

phosphate insecticide was incorporated into panels of polychlorovinyl resin, which permitted the insecticide to volatilize at a much reduced rate, killing insects in the immediate vicinity. The most recent slow-release formulation
appeared in 1984, a microencapsulated concentrate of permethrin (Penncapthrin-200[R]) for use in fly control.

The principle of this slow-release formulation involves
the incorporation of the insecticide in a permeable covering, microcapsules or tiny spheres, with diameters ranging
from 30 to 50 μm, that permits its release at a reduced,
but effective rate. The insecticide escapes through the
sphere wall over an extended period, thus preserving its
effectiveness much longer, usually from 2 to 4 times, than
if formulated as an emulsifiable concentrate. In general
the slow release formulations increase the residual life
of the insecticide and thus reduce the number of applications needed. Panel E details those slow-release formulations currently available.

PANEL E
SLOW-RELEASE INSECTICIDES

Product, Insecticide and Manufacturer	Physical Form	Use	Method of Application
No-Pest Strips[R], DDVP (20%) Texize, Division Morton-Norwich	Resin strip	Fumigant for flying insects	Hung near ceiling
Insect Strip[R], DDVP (20%) Starbar, Division of Zoecon Corp.	Resin strip	Fumigant for flying insects	Hung near ceiling
Killmaster[R] II, chlorpyrifos (2%) Positive Formulators, Inc.	Ready-to-use liquid	Crawling house-hold pests	Professional application only
Killmaster[R] 6 Months Pest Control[R] and Positive Control[R], chlorpyrifos (1%) Positive Formulators, Inc.	Ready-to-use liquid	Crawling house-hold pests	Homeowner application by brush or coarse spray
Penncap M[R], methyl parathion (22%) Pennwalt Corp.	Microencapsulated concentrate (30-50 μm dia.)[a]	Agricultural pests	Professional application only
Penncap E[R], ethyl parathion (22%) Pennwalt Corp.	Microencapsulated concentrate (30-50 μm dia.)	Agricultural pests	Professional application only

[a] An average human hair is 50 μm in diameter.

PANEL E (Continued)

Product, Insecticide and Manufacturer	Physical Form	Use	Method of Application
Penncapthrin-200[R], permethrin (20%) Pennwalt Corp.	Microencapsulated concentrate (30-50 μm dia.)	Residual spray for fly control	Not registered in U. S.
Knox Out 2FM[R], diazinon (20%) Pennwalt Corp.	Microencapsulated concentrate (30-50 μm	Agricultural pests and crawling household pests	Professional application only for agriculture, and household
Sectrol #90 Concentrate[R] pyrethrins + synergists (1.1% + 5.9%) 3M, Industrial Tape Div.	Microencapsulated concentrate (15-20 μm dia.)	Crawling and flying household insects	Professional application only as spray or ULV aerosol
Hercon Insectape[R] With Chlorpyrifos (10%) With Diazinon (10%) With Propoxur (10%) Hercon Div., Health-Chem Corp.	Ready-to-use, adhesive, laminated strips	Crawling household pests	Tape applied to surfaces
Roach-Tape[R], propoxur (10%) Hercon Div., Health-Chem Corp.	Ready-to-use adhesive, laminated strips	Crawling household pests	Tape applied to surfaces
Altosid Briquets[R], 4% methoprene Professional Pest Management Div., Zoecon Corp.	Ready-to-use briquets, 68 grams each	Mosquito growth regulator prevents adult mosquito emergence	Briquets placed in mosquito breeding waters

Killmaster[R], the paint-on slow-release formulation containing chlorpyrifos, is a recent innovation in structural pest control. The insecticide is dissolved in a volatile petroleum solvent containing a unique combination of dissolved plastics and lacquers in small quantities. Following its paint-on application as a spot treatment in homes, restaurants, and food-handling establishments, the solvent quickly evaporates, leaving the insecticide incorporated in a thin transparent film. Over time the insecticide "blooms" or escapes to the surface at a constant rate presenting at all times a freshly exposed surface to crawling insects. Insecticidal adhesive tapes work as contact insecticides against crawling insects. The toxicant is laminated into the multilayered, adhesive polymeric strips, attached beneath counters, under shelves, and in other protected places. These two slow-release formulations are now available under various brand names to homeowners.

4.23 Two reasons for formulating a pesticide in a slow-release form are _____and
_____.

- -

Reduce number of applications
Increase life of highly volatile material

4.24 What are some of the potential problems that could result from the use of non-formulated, technical grade, pesticides?

_____
_____
_____
_____

- -

Hazard to the applicator; Danger to bystander, domestic animals, and pets; Contamination of food and water supplies; Excessive cost; Difficulties in application; Excessive residual life; Poor storage qualities and shelf-life; Toxicity to plants (phytotoxicity); Reduced effectiveness; and many others.

It might appear that there is no limit to the different forms in which a pesticide can be prepared. This is almost the case. If not through economy, then by EPA restrictions, we will learn to formulate and apply pesticides in extremely conservative ways, to preserve our health, our resources, and the environment. In summary, pesticides are formulated to improve their properties of storage, handling, application, effectiveness and safety.

UNIT 5
INSECTICIDES

Man has been on the earth for about one million years. Insects have survived for more than 250 million years! We don't know, but the first materials used by primitive man, classed as insecticides (repellents) in the crudest definition of the word, were probably mud and dust spread over his skin to repel biting insects.

The earliest records of insecticides pertain to the burning of "brimstone" (sulfur) as a fumigant. Pliny the Elder (A.D. 23-79) recorded most of the earlier insecticide uses in his Natural History, collected largely from the folklore and Greek writings of the previous two or three centuries. Included among these were the use of gall from a green lizard to protect apples from worms and rot. In the interim a variety of materials have been used with questionable results: Extracts of pepper and tobacco, hot water, whitewash, vinegar, turpentine, fish oil, brine, lye and many others.

In 1944, our insecticide supply was still limited to several arsenicals, petroleum oils, nicotine, pyrethrum, rotenone, sulfur, hydrogen cyanide gas, and cryolite. And then, World War II opened the Chemical Era with the introduction of a totally new concept of insect control chemicals, synthetic organic insecticides, the first of which was DDT.

5-A ORGANOCHLORINES

The organochlorines are insecticides that contain carbon (thus the term organo-), chlorine, hydrogen, and sometimes oxygen. They are also referred to by other names as "chlorinated hydrocarbons," "chlorinated organics," "chlorinated insecticides," and "chlorinated synthetics."

5.1 Organochlorine insecticides contain the elements
_____, _____, and _____.

- -

carbon, chlorine, hydrogen (and sometimes oxygen)

5.2 Two synonyms for the organochlorines are
_____ and _____.

- -

      Chlorinated hydrocarbons, or chlorinated organics, or
      chlorinated insecticides, or chlorinated synthetics

      DDT and Relatives. The best known and the most notor-
ious of this group, is DDT, which probably will be declared
the most useful insecticide known. If only for historical
purposes, let's begin with the fascinating story of DDT.
      DDT, surprisingly, is more than 100 years old. It was
first synthesized by a German graduate student in 1873,
with no idea of its tremendous insecticidal value and,
after synthesis, it was thrown out and forgotten. In 1939
a Swiss entomologist, Dr. Paul Müller, rediscovered DDT
while searching for a long-lasting insecticide against the
clothes moth. DDT proved to be extremely effective against
flies and mosquitoes, ultimately bringing to Dr. Müller
the Nobel Prize for Medicine in 1948 for his life-saving
discovery.
      More than 4 billion pounds of DDT have been used
throughout the world for insect control since 1940, and
80% of that amount was used in agriculture. Production
reached its maximum in the U. S. in 1961, when 160 million
pounds were manufactured. The greatest agricultural benefits from DDT were in the control of the Colorado potato
beetle and several other potato insects, the codling moth
on apples, corn earworm, cotton bollworm, tobacco budworm,
pink bollworm on cotton, and the worm complex on vegetables.
It was most useful against the gypsy moth and the spruce
budworm in forests. From the standpoint of human medicine,
DDT was and still is in many areas, highly successful
against mosquitoes that transmit malaria and yellow fever,
against body lice that can carry typhus, and against fleas
that are vectors of plague.
      One of the amazing features of DDT was its low cost.
Most of that sold to the World Health Organization went for
less than 22 cents per pound! Without question, it was the
most economical insecticide ever sold. A federal ban on
the use of DDT was declared by the Environmental Protection
Agency on January 1, 1973. DDT was declared to be an environmental hazard due to its long residual life and its accumulation, along with the metabolite DDE, in food chains
where it proved to be detrimental to certain forms of wildlife.
      DDT belongs to the chemical class of *diphenyl ali-
phatics,* meaning that it consists of an aliphatic, or
straight carbon chain, with two (di-) phenyl rings attached
as in the illustrations.

DDT was first known chemically as dichloro diphenyl trichloroethane, thus ddt or DDT.

DDT

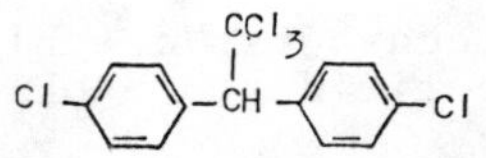

1,1,1-trichloro-2,2-bis(p-chlorophenyl)ethane

There are three relatives of DDT which are still used. Methoxychlor is a good home, yard and garden insecticide and is now used as an impregnant of woolen articles at the cleaning plant for moth-proofing. Dicofol and chloro-benzilate are both miticides. Ethylan was formerly used as the moth-proofing impregnant of woolens, however all uses voluntarily cancelled by the manufacturer.

METHOXYCHLOR

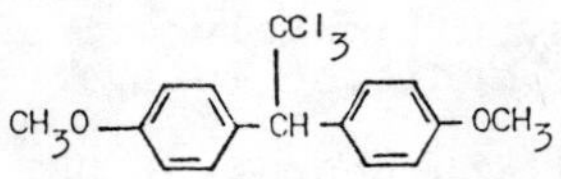

1,1,1-trichloro-2,2-bis(p-methoxyphenyl)ethane

DICOFOL(Kelthane^R)

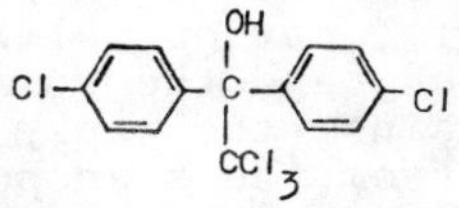

4,4'-dichloro-α(trichloromethyl)benzhydrol

CHLOROBENZILATE

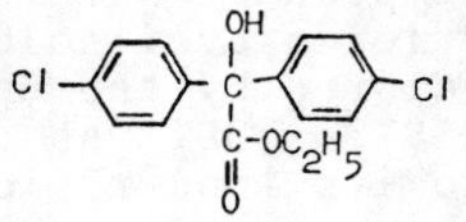

ethyl 4,4'-dichlorobenzilate

ETHYLAN (Perthane[R])

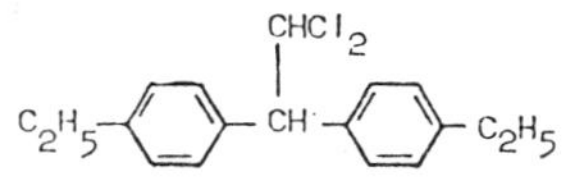

$$CHCl_2$$

$$C_2H_5 - \langle \rangle - CH - \langle \rangle - C_2H_5$$

1,1-dichloro-2,2-bis(p-ethylphenyl)ethane

The structures really don't reveal anything about their chemical stability, or *persistence*, but only DDT and TDE (not illustrated) have this characteristic. Persistence, as used here, implies a chemical stability giving the products long lives in soil and aquatic environments, and in animal and plant tissues. They are not readily broken down by microorganisms, enzymes, heat, or ultraviolet light. From the insecticidal viewpoint these are good characteristics. From the environmental viewpoint they are not. Using these criteria, the DDT relatives illustrated would be considered non-persistent.

5.3 Of the diphenyl aliphatics in the class of organo-
     chlorine insecticides, only _____ and _____
     are considered persistent.

- -

     DDT   TDE

5.4 What are the characteristics of a persistent insec-
     ticide?

     _____
     _____
     _____

- -

     chemical stability; long life in soil, aquatic envir-
     onment, plant and animal tissues; not readily broken
     down by microorganisms, enzymes, heat or UV

The *mode of action*, or type of biological activity, has never been clearly worked out for DDT or any of its relatives. It does affect the neurons or nerve fibers in a way that prevents normal transmission of nerve impulses, both in insects and mammals. Eventually the neurons fire impulses spontaneously, causing the muscles to twitch; this may lead to convulsions and death.

DDT does not seem to react with any particular enzyme, thus it is theorized that the physical interference of DDT with membrane permeability is an important factor in its effect. The prevailing theory suggests that DDT acts as a "wedge in the outer covering of the neuron". In this fashion DDT molecules keep the sodium gates open and leaking so that the membrane resting potential, or differential voltage, cannot be restored. This causes the repeated discharge resulting in tetanus. It is sufficient to know that DDT and its relatives in a complex way destroy the delicate balance of sodium and potassium around the neuron, giving it a "short circuit", and so prevent it from conducting impulses normally.

5.5   DDT acts by upseting the balance of _____ and _____ in the affected neuron.

- -

sodium and potassium

Benzene Derivatives. Benzene is a six-membered carbon ring, having the structure in the opposite figure. It is the building block for a few insecticides, some of which were used in great quantity at one time. The first was BHC (HCH), benzenehexachloride, first reported in 1825. But like DDT, was not known to have insecticidal properties until 1940, when French and British entomologists found the material to be active against all insects tested.

BENZENE

C_6H_6

BHC is nothing more than chlorinated benzene, which results in a product made up of several isomers, that is, molecules containing the same kinds and numbers of atoms but differing in the internal arrangement of those atoms. BHC, for instance, has 5 isomers named after the Greek letters, Alpha, Beta, Gamma, Delta and Epsilon. After much laboratory work in isolating and identifying these isomers, the chemists found to their great surprise that only the *Gamma isomer* had insecticidal properties. In a normal mixture of BHC, the Gamma isomer comprises only about 12% of the total, leaving the other four isomers as inert material, or insecticidally inactive ingredients.

5.6   The only insecticidal isomer of BHC is the _____ isomer. Of what value are the remaining isomers?

- -

gamma   Of no value as insecticides

BHC (HCH)

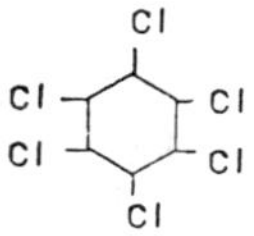

CI
CI ⎯ ⎯ CI
CI ⎯ ⎯ CI
CI

1,2,3,4,5,6-hexachlorocyclohexane

Because the Gamma isomer was the active ingredient, methods were developed to manufacture a product containing 99% Gamma isomer, Lindane, which was effective against most insects, but also quite expensive, making it impractical for crop use.

BHC had one highly undesirable characteristic, a prominent musty odor and flavor, which helps in remembering several other points of interest. Apparently this odor is emitted by the inert isomers, which are much more persistent than the odorless Gamma isomer in animal and plant tissues as well as in soil. As a result, root and tuber crops planted in soils previously treated with BHC retained the odor of BHC and were unsalable. There were similar reports of this problem with leafy vegetables, poultry, eggs, and milk, which came in contact with BHC residues.

5.7   BHC's most undesirable characteristic was the persistent _____ left in plant and animal tissues exposed to it.

- -

    odor, flavor

Lindane's mode of action is also not completely understood. Its biological effects on insects and mammals are definitely different from those of DDT, but are also probably brought on by a sodium-potassium imbalance in the neurons. It is known to be a neurotoxicant whose effects are normally seen much faster than DDT, and result in increased activity, tremors, and convulsions leading to prostration. Lindane has few remaining uses, one of which is in the form of lotions prescribed by physicians for control of head, body and crab lice.

5.8 Lindane's toxicity is exerted against the _____.

----

    nerves or neurons

    Cyclodienes. This prominent and extremely useful
group of insecticides is also known as the dieneorgano-
chlorine insecticides. First, let's see why they are
named cyclodienes. *Cyclo* obviously means cyclic or ring
structures, and *-diene* means "containing two double bonds,"
such as the ring structure shown.

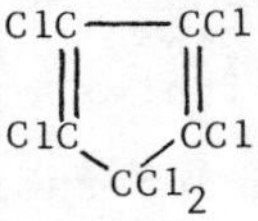

$$
\begin{array}{ccc}
ClC & \!\!-\!\!-\!\! & CCl \\
\| & & \| \\
ClC & & CCl \\
 & \diagdown \quad \diagup & \\
 & CCl_2 &
\end{array}
$$

    This molecule is hexachlorocyclopentadiene, a build-
ing block or *precursor* for most insecticides in this
group.   These cyclodienes are joined to other similar compounds in diene-synthesis, the Diels-Alder reaction, named
after two German chemists who described this chemical
reaction.   Two well-known insecticides, aldrin and dieldrin,
belong to this group of cyclodienes and were named after
these Nobel prize-winning chemists.

5.9 Cyclodienes are insecticides synthesized by the
    _____ reaction from building blocks, or
    _____, usually one of which is
    hexachlorocyclopentadiene.

    Diels-Alder   precursors

    The chemistry and nomenclature of the cyclodienes is
rather complicated and will not be studied in depth.   It
is of value to mention that the cyclodienes do have
3-dimensional structures, and thus possess *stereoisomers,*
whose atoms differ in their spatial location and structure.
In the cyclodienes there is one and frequently two methano-
bridges, one located in the chlorinated ring and the other
in the unchlorinated ring.

5.10 In this typical cyclodiene structure identify the
moieties indicated by (1) _____
and (2) _____ .

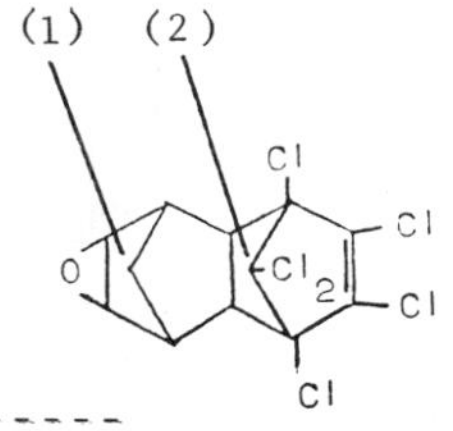

(1) methano-bridge in the unchlorinated ring
(2) methano-bridge in the chlorinated ring

Because the methano-bridges are bent "inside" (*endo-*)
or "outside" (*exo-*) of the cage structure they are used in
the chemical nomenclature to describe the stereochemistry
of the molecules, as in the case of dieldrin and endrin.

DIELDRIN

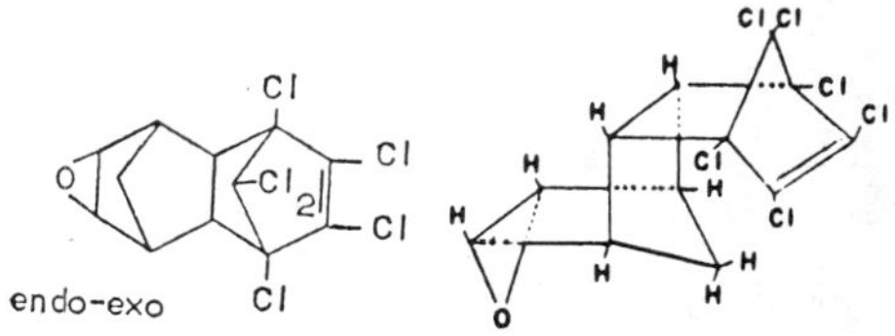

endo-exo

1,2,3,4,10,10-hexachloro-6,7-epoxy-1,4,4a,5,6,7,8,8a-
octahydro-1,4-endo-exo-5,8-dimethanonaphthalene

ENDRIN

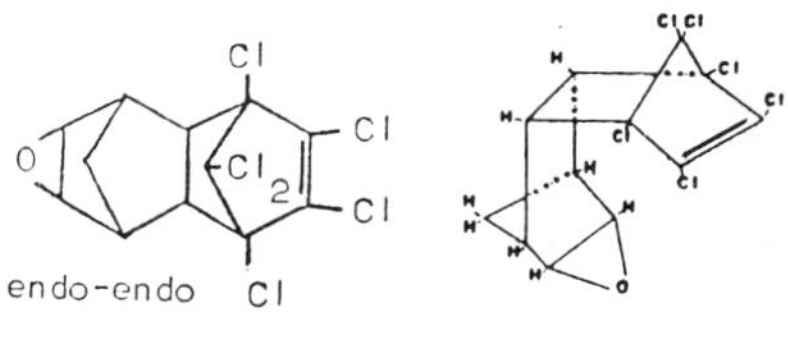

endo-endo

1,2,3,4,10,10-hexachloro-6,7-epoxy-1,4,4a,5,6,7,8,8a-
octahydro-1,4-endo-endo-5,8-dimethanonaphthalene

48   INSECTICIDES

The cyclodienes were developed after World War II,
and are therefore of more recent origin than DDT (1939)
and BHC (1940). The 8 compounds listed below were first
described in the scientific literature or patented in the
year indicated: chlordane, 1945; aldrin and dieldrin,
1948; heptachlor, 1949; endrin, 1951; mirex, 1954; endosulfan, 1956; and chlordecone, 1958. Other cyclodienes
developed in the United States and Germany included Isodrin,
Alodan, Bromodan, and Telodrin, which are no longer registered in the U. S.

Generally, the cyclodienes are persistent insecticides
and are stable in soil and relatively stable to the action
of sunlight. Consequently, they were used in greatest quantity as soil insecticides, especially chlordane, heptachlor,
aldrin and dieldrin, for the control of termites and soil-
borne insects whose immature stages (larvae) feed on the
roots of plants. Endosulfan is the only cyclodiene still
registered for use on food drops.

To suggest the effectiveness of cyclodienes as termite
control agents, structures treated with chlordane, aldrin,
and dieldrin in the year of their development are still protected from damage. This is 41 and 38 years, respectively.
It would be elementary to say that these insecticides are
the most effective, long-lasting, economical, and safest
termite control agents known. However, several other soil
insects became resistant to these materials in agriculture,
resulting in a rapid decline of their use. Most agricultural uses of the cyclodienes were cancelled by EPA between
1975 and 1980.

5.11 The outstanding characteristic of the cyclodienes is
     their _____, which made them ideal
     for the control of _____ insects.

- -

     persistence or stability   soil

CHLORDANE

1,2,4,5,6,7,8,8-octachloro-3a,4,7,7a-tetrahydro-4,7-methanoindane

HEPTACHLOR

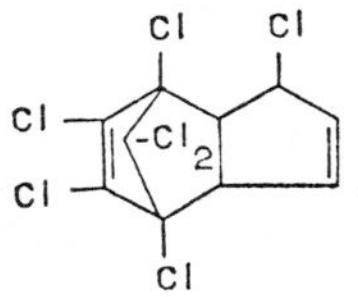

1,4,5,6,7,8,8-heptachloro-3a,4,7,7a-tetrahydro-4,7-methanoindene

ALDRIN

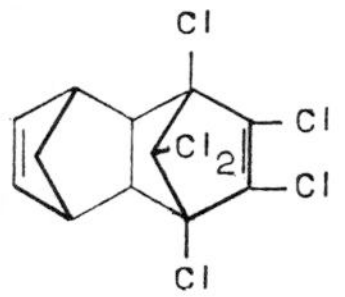

1,2,3,4,10,10-hexachloro-1,4,4a,5,8,8a-hexahydro-
1,4-endo-exo-5,8-dimethanonaphthalene

ENDOSULFAN (Thiodan®)

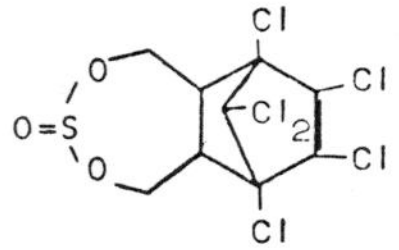

6,7,8,9,10,10-hexachloro-1,5,5a,6,9,9a-hexahydro-6,9-
methano-2,4,3-benzodioxathiepin 3-oxide

MIREX

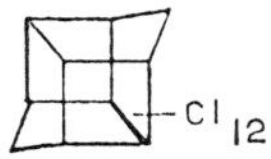

dodecachlorooctahydro-1,3,4-metheno-1H-cyclobuta[cd]pentalene

CHLORDECONE (Kepone^R)

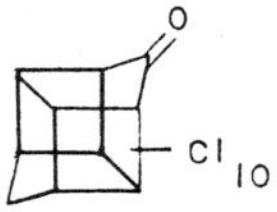

decachlorooctahydro-1,3,4-metheno-2H-cyclobuta[cd]pentalen-2-one

The cyclodienes are generally *equitoxic*; they have equal toxicity or toxic effects, to insects, mammals, and birds. That is, given the same dosage based on weight, such as milligrams per kilogram (mg/kg) of body weight, these materials have about the same degree of toxicity. There are always exceptions, and fish are much more susceptible perhaps for a very good reason - they are totally surrounded when the compound is introduced into water. Figuratively speaking, they eat, sleep, and breathe the toxicant in their aquatic environment.

5.12 A material having equal toxicity to different species of animals on a weight/weight basis is said to be _____ to these animals.

- -

equitoxic

5.13 Why are fish so sensitive to the toxic effects of the organochlorine insecticides?

- -

Fish eat, breathe, and are surrounded by the toxicant plus the fact that the organochlorines are generally the most chemically stable of the insecticides.

The mode of action of this class of organochlorine insecticides is also not clearly understood. It is known that they are neurotoxicants, having effects very similar to those of DDT. They appear to affect all animals in generally the same way, first with nervous activity followed by tremors, convulsions, and prostration. The cyclodienes undoubtedly molest the delicate balance of sodium and potassium within the neuron. It is quite likely that the mechanism of cyclodiene poisoning also involves changes in ion permeability of axonic membranes, as with DDT.

5.14 Cyclodienes act as _____ probably upsetting the balance of _____ and _____ ions.

- -

neurotoxicants   sodium and potassium

Polychloroterpene Insecticides. Only two materials,
toxaphene discovered in 1947, and strobane, introduced in
1951, belong to this group. Neither were used in urban
pest control. Toxaphene was manufactured by the chlorina-
tion of camphene, a pine tree derivative, to yield a
material containing 67-69% chlorine by weight, with the
molecular formula $C_{10}H_{10}Cl_8$. Strobane was produced by
chlorinating alpha-pinene, yielding a 66% chlorinated
material with the molecular formula of $C_{10}H_{11}Cl_7$.
Toxaphene had by far the greatest use of any single
insecticide in agriculture. It was used mostly on cotton
in combination with DDT until around 1965. After most
cotton insects became resistant to DDT toxaphene was then
formulated with methyl parathion, an organophosphate insecticide.
Most registered uses of toxaphene were cancelled in
1983 by the EPA. However, sale and distribution of existing stocks are permitted through December, 1986.
There are more than 177 polychlorinated derivatives
(different kinds of molecules) in toxaphene. No single
component makes up more than a small percent of the product.
These materials are persistent in soil, but not to the
extent of the cyclodienes. They disappear in 3 to 4 weeks
from most plant surfaces. This disappearance is due primarily to volatility and not to metabolism or photolysis
(breakdown by ultraviolet). They are metabolized by mammals and also stored in body fat, but not to the extent
that DDT or the cyclodienes are stored. Despite its low
toxicity to insects, mammals and birds, fish are highly
susceptible to toxaphene, as they are to the cyclodiene
insecticides.
The mode of action for toxaphene is similar to the
cyclodienes, acting on the neurons and causing a sodium
and potassium ion imbalance.

TOXAPHENE

chlorinated camphene containing 67-69% chlorine

STROBANE

$C_{10}H_{11}Cl_7$

terpene polychlorinates (65% chlorine)

5.15 The polychloroterpene insecticides are considered
     _____ materials.
- -

     persistent

_____

5.16 Toxaphene and strobane have modes of action similar
     to the cyclodiene insecticides. They act on the
     _____ causing an imbalance in _____
     and _____ ions.
- -

     neurons   sodium   potassium

_____

5.17 List several advantages and disadvantages of persis-
     tent insecticides.
     Advantages:_____
     Disadvantages:_____
- -

     Advantages: Low cost, effective for long periods,
                 good soil insecticides.
     Disadvantages: Build up in food chains, residues
                    don't break down quickly; very toxic
                    to all aquatic life; stored in tissues
                    of man and other animals.

_____

5.18 Place the appropriate molecular structure number
     after the organochlorine subclass in which it
     belongs: DDT relatives_____; polychloroterpenes____;
     cyclodienes_____; and benzene derivatives _____.

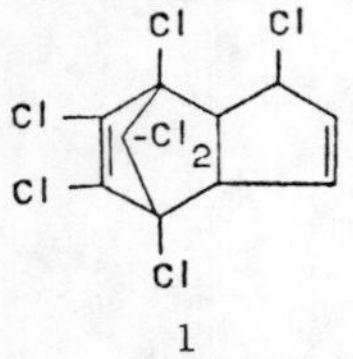

1

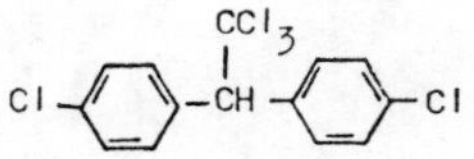

2

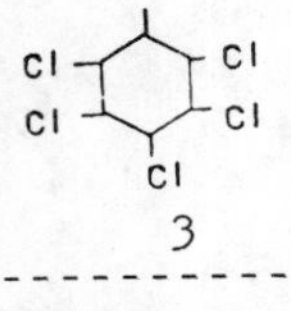

3

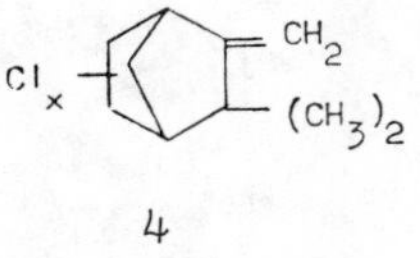

4

- -

     DDT relatives   2   polychloroterpenes   4
     cyclodienes   1   benzene derivatives   3

5-B ORGANOPHOSPHATES

The chemically unstable organophosphate insecticides
(a generic term which includes all insecticides containing
phosphorus) have several commonly used names: organic
phosphates, phosphorus insecticides, nerve gas relatives,
phosphate insecticides, and phosphorus esters or phosphoric acid esters.

$$\text{O} \atop \text{||}$$

They are all derived from phosphoric acid, HO-P-OH, and

$$\text{O} \atop \text{H}$$

generally the most toxic of all pesticides to vertebrate
animals. Indeed, they are related to the "nerve gases"
by chemical structure and mode of action. Their similari-
ties end here, for their toxicities are in no way alike
being several magnitudes apart.

5.19 Two synonyms for the organophosphate insecticides
     are _____ and _____ .

- -

     organic phosphates, phosphate insecticides, phos-
     phorus insecticides, phosphates, etc.

The insecticidal action of these compounds was discovered in Germany during World War II in the study of
materials closely related to the nerve gases, sarin,
soman, and tabun. Initially, the discovery was made in
search for substitutes for nicotine, which was used as an
insecticide and in critically short supply in Germany.

5.20 The organophosphate insecticides are derivatives of
     _____ acid and are related to the
     _____ by their chemistry and
     mode of action.

- -

     phosphoric   nerve gases

The organophosphates (OPs) have two distinctive
features. First, they are generally much more acutely toxic
to vertebrates than the organochlorine insecticides, and
second, they are non-persistent. It is this latter quality which brought them onto the agricultural scene to
gradually displace the persistent organochlorines, particularly DDT.

5.21 Organophosphates differ from organochlorines by being
     much more _____ to vertebrates and less
     _____.

- -

     toxic   persistent

    Before we delve deeper into this gigantic group of
compounds, we can discuss their mode of action with considerable confidence. The OPs exert their toxic action by
tying up or *inhibiting* certain important enzymes of the
nervous system, *cholinesterases* (ChE). Throughout the
nervous system in vertebrates as well as insects, are
electrical switching centers, or *synapses*, where the electrical signal is carried across a gap to a muscle or
another nerve fiber (neuron) by a chemical, in many instances, acetylcholine (ACh). After the electrical signal
(nerve impulse) has been conducted across the gap by ACh,
the ChE enzyme moves in quickly and removes the ACh so the
circuit won't be "jammed." These chemical reactions hap-
pen extremely rapidly and go on constantly under normal
conditions. When OPs enter the scene, they attach to the
ChE in a way that prevents them from removing the ACh.
The circuits then jam because of the accumulation of ACh.
What this really says is that the accumulation of ACh
interferes with the neuromuscular junction, giving rise to
rapid twitching of voluntary muscles and finally to paralysis. This is of particular importance in proper func-
tioning of the respiratory system.
    The mode of action of organophosphates is not quite
as simple as outlined above, but as presented serves the
purposes of this general study.

5.22 Organophosphates, abbreviated to _____, act by
     inhibiting enzymes known as _____,
     and abbreviated to _____.

- -

     OPs   cholinesterases   ChE

5.23 The switching centers in the nervous system are
     referred to as _____.

- -

     synapses

5.24 The nerve impulse is transmitted across this synapse
     by the chemical _____abbreviated as
     _____.

- -

     acetylcholine   ACh

5.25 When OPs inhibit ChE there is an accumulation of
     _____ at the _____.

- -

     acetylcholine   synapse

     We will now consider the atoms attached to the phos-
phorus. OPs that are combinations of different alcohols
and different phosphorus acids are termed *esters*.

5.26 Esters are the result of combining an _____
     with an _____. Thus OPs are esters of
     _____acids and an _____.

- -

     acid   alcohol   phosphorus   alcohol

     Esters of phosphorus have varying combinations of
oxygen, carbon, sulfur, and nitrogen attached to the phosphorus, and so have different identities. Below are shown
the six sub-classes of OPs, only to help explain some of
the seemingly odd chemical names given to the materials
studied.

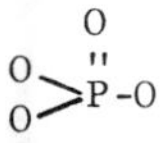

Phosphate

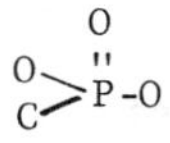

Phosphonate

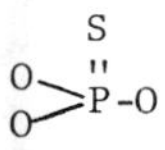

Phosphorothioate

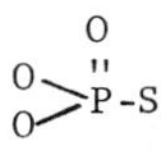

Phosphorothiolate

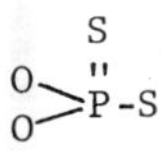

Phosphorodithioate

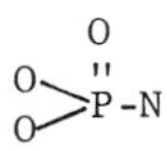

Phosphoramidate

5.27 Name four of the six subclasses of OPs.

_____   _____

_____   _____

Phosphate   Phosphonate   Phosphorothioate
Phosphorothiolate Phosphorodithioate   Phosphoramidate

As with the last two groups of insecticides, the OPs
are further divided into three classes, the aliphatic,
phenyl, and heterocyclic derivatives. Each class has
several materials to be examined.

Aliphatic Derivatives. *Aliphatic* literally means
"carbon chain," and the linear arrangement of carbon atoms
differentiates them from ring or cyclic structures. All
of the aliphatic OPs are simple phosphoric acid derivatives bearing short carbon chains.

5.28 Aliphatics are _____ _____.

----

carbon chains

The oldest and most heavily used insecticide of this
classification is malathion. Introduced in 1950, it was
quickly adopted by agriculture for use on most vegetables,
fruits, and forage crops for control of an extensive range
of insect pests. It was soon recognized as the answer to
the homeowner's prayer, for it was safe to use around pets,
fast-acting, and controlled practically every kind of garden and household insect including aphids and cockroaches.
It is so safe that it is prescribed by physicians for use
on humans for the control of head, body and crab lice. It
commonly appears in flea powders for dogs, cats, and other
domestic animals, and is used in dips for the control of
mange mites.

MALATHION (Cythion[R])

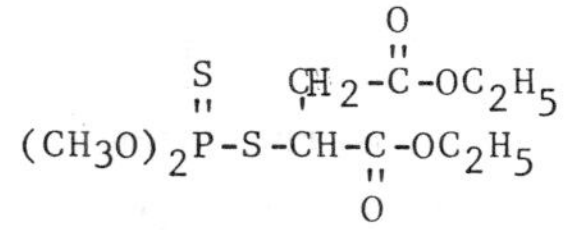

$$\begin{array}{cc} & O \\ S & CH_2\text{-}C\text{-}OC_2H_5 \\ \| & \\ (CH_3O)_2P\text{-}S\text{-}CH\text{-}C\text{-}OC_2H_5 \\ & \| \\ & O \end{array}$$

O,O-dimethyl-S-1,2-di(carboethoxy)
ethyl phosphorodithioate

In 1981 it became the insecticide of choice in the control of the Mediterranean fruit fly found in the rich fruit growing areas of California. Malathion was mixed with a protein bait made of molasses and yeast and sprayed either from ground equipment or by helicopter over the infested and surrounding areas. Using this technique, malathion was applied at the astonishingly low rate of 2.5 ounces per acre mixed with 10 ounces of bait. Both male and female fruit flies are attracted to the bait and die a few hours after feeding.

Trichlorfon is a chlorinated OP, which has been useful for crop pest control and fly control around barns and other farm buildings.

TRICHLORFON (Dipterex[R], Dylox[R])

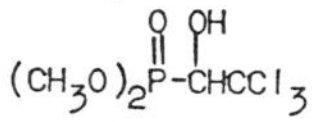

$$\begin{array}{cc} O & OH \\ \| & | \\ (CH_3O)_2P\text{-}CHCCl_3 \end{array}$$

dimethyl (2,2,2-trichloro-1-hydroxyethyl)phosphonate

Naled is a widely used brominated OP with low mammalian toxicity. It is registered for many crops with a waiting interval of 4 days or less, and for fly control in barns, poultry houses, kennels, horse corrals and around food processing plants. Because of its fumigant quality it can be applied to the pipes of greenhouse heating systems to kill insects by vapor action. It is also widely applied in large area mosquito control programs by ULV or conventional sprays.

NALED (Dibrom[R])

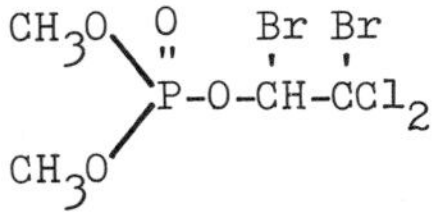

$$\begin{array}{c} CH_3O \\ \quad\quad O \quad Br \ Br \\ \quad\quad \| \quad | \ | \\ \quad\quad P\text{-}O\text{-}CH\text{-}CCl_2 \\ CH_3O \end{array}$$

1,2-dibromo-2,2-dichloroethyl-
dimethyl phosphate

Monocrotophos is an aliphatic OP containing nitrogen. It is a plant systemic insecticide, but it has had limited use in agriculture because of its high mammalian toxicity. It is also not available to the homeowner.

MONOCROTOPHOS (Azodrin®)

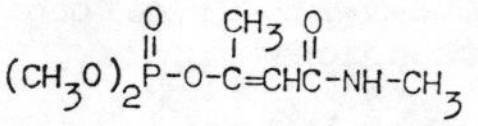

$$(CH_3O)_2\overset{O}{\overset{\|}{P}}-O-\overset{CH_3}{\overset{|}{C}}=\overset{O}{\overset{\|}{CHC}}-NH-CH_3$$

3-hydroxy-N-methyl-cis-crotonamide dimethyl phosphate

Plant systemic insecticides are those that are taken into the roots and translocated to the above-ground parts, where they are toxic to any sucking insects feeding on the plant juices. Normally caterpillars and other plant tissue-feeding insects are not controlled, because they do not ingest enough of the systemic-containing juices to be affected.

_____

5.29 Systemic insecticides are useful only against _____ insects.

- -

sucking

_____

Among the aliphatic derivatives are several plant systemics: Dicrotophos, demeton, phorate, dimethoate, oxydemetonmethyl, and disulfoton. Only the latter three have formulations which are available for use by the homeowner.

DIMETHOATE (Cygon®)

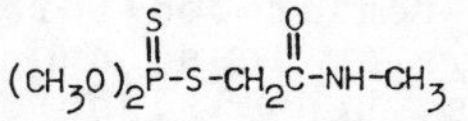

$$(CH_3O)_2\overset{S}{\overset{\|}{P}}-S-CH_2\overset{O}{\overset{\|}{C}}-NH-CH_3$$

O,O-dimethyl S-(N-methylcarbamoylmethyl)phosphorodithioate

DICROTOPHOS (Bidrin^R)

$$(CH_3O)_2\overset{O}{\overset{\|}{P}}-O-\overset{CH_3}{\overset{|}{C}}=\overset{O}{\overset{\|}{CHC}}-N(CH_3)_2$$

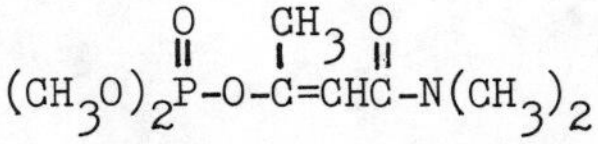

O,O-dimethyl O-1-methylvinyl-N,N-dimethyl carbamoyl phosphate

OXYDEMETON-METHYL (Metasystox-R^R)

$$(CH_3O)_2\overset{O}{\overset{\|}{P}}-S-CH_2CH_2-\overset{O}{\overset{\|}{S}}-CH_2CH_3$$

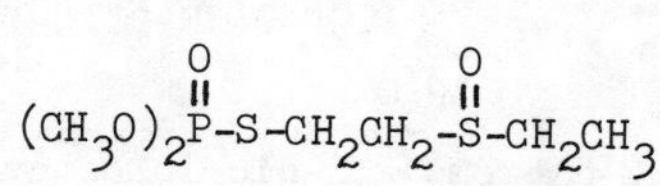

S-(2-(ethylsulfinyl)ethyl)O,O-dimethyl phosphorothioate

DISULFOTON (Di-Syston[R])

$$(C_2H_5O)_2\overset{\overset{S}{\|}}{P}-S-CH_2CH_2-S-C_2H_5$$

O,O-diethyl S-2-[(ethylthio)ethyl]phosphorodithioate

DEMETON (Systox[R])

$$(C_2H_5O)_2\overset{\overset{O}{\|}}{P}-S-CH_2CH_2-S-CH_2CH_3$$

mixture of

$$(C_2H_5O)_2\overset{\overset{S}{\|}}{P}-O-CH_2CH_2-S-CH_2CH_3$$

mixture of O,O-diethyl S-(and O)-2-[(ethylthio)ethyl]phosphorothioates

PHORATE (Thimet[R])

$$(C_2H_5O)_2\overset{\overset{S}{\|}}{P}-S-CH_2-S-C_2H_5$$

O,O-diethyl-S-[(ethylthio)methyl] phosphorodithioate

Dichlorvos is an aliphatic OP with a very high vapor pressure, giving it strong fumigant qualities. It has been incorporated into polychlorovinyl resin pest strips and pet collars from which it is released slowly. It lasts several months, and is useful for insect control in the home and other closed areas.

DICHLORVOS (DDVP, Vapona[R])

$$(CH_3O)_2\overset{\overset{O}{\|}}{P}-O-CH=CCl_2$$

O,O-dimethyl-O-2,2-dichloro-
vinyl phosphate

Mevinphos is a highly-toxic OP used in vegetable production because of its very short residual life. It can be applied up to one day before harvest for insect control, yet it leaves no residue on the crop to be eaten by the consumer.

MEVINPHOS (Phosdrin[R])

$$(CH_3O)_2\overset{\overset{O}{\|}}{P}-O-\underset{\underset{CH_3}{|}}{C}=CH\overset{\overset{O}{\|}}{C}-OCH_3$$

2-methoxy-carbonyl-1-methylvinyl dimethyl phosphate

Monitor[R] and Orthene[R] are two of the recent arrivals in the aliphatic organophosphates. Both have proved highly useful in agriculture, especially for vegetable insect control. Terbufos is formulated as a granular for the control of corn rootworm and other soil insects and

nematodes infesting field corn, as well as maggots in
sugar beets.

METHAMIDOPHOS (Monitor[R])

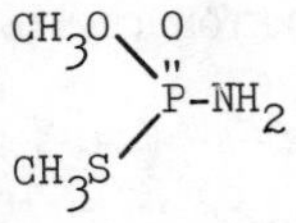

$$CH_3O \diagdown \underset{CH_3S \diagup}{\overset{O}{\underset{\|}{P}}} -NH_2$$

O,S-dimethyl phosphoramidothioate

ACEPHATE (Orthene[R])

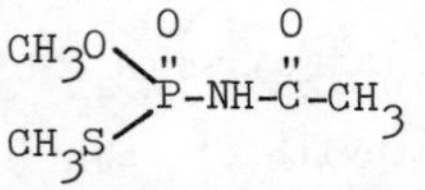

$$CH_3O \diagdown \underset{CH_3S \diagup}{\overset{O}{\underset{\|}{P}}} -NH-\overset{O}{\underset{\|}{C}}-CH_3$$

O,S-dimethyl acetylphosphoramidothioate

TERBUFOS (Counter[R])

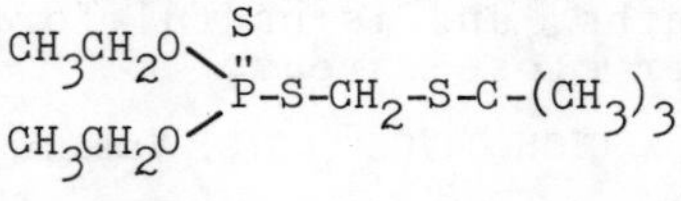

$$CH_3CH_2O \diagdown \underset{CH_3CH_2O \diagup}{\overset{S}{\underset{\|}{P}}} -S-CH_2-S-C-(CH_3)_3$$

S-(((1,1-dimethylethyl)thio)methyl)
O,O-diethyl phosphorodithioate

    In summary, the aliphatic organophosphate insecticides
are the simplest structures of the organophosphate molecules. They have a wide range of toxicities, and several
possess a relatively high water solubility, giving them
plant-systemic qualities several of which are useful around
the home. Because of their acute toxicities, many of the
aliphatic OPs are restricted use pesticides.

    Phenyl Derivatives. The phenyl OPs contain a benzene
ring with one of the ring hydrogens displaced by attachment to the phosphorus moiety and others frequently displaced by Cl, NO_2, CH_3, CN, S, and so on. The phenyl OPs
are generally more stable than the aliphatic OPs; consequently their residues are longer lasting.
    Parathion is the most familiar of the phenyl OPs.
It was introduced in 1947, the second organophosphate used
in agriculture (TEPP appeared in 1946). Because of its age
and wide utility, its total usage is greater than that of
many of the less useful materials combined. Ethyl parathion,

its proper name, was the first phenyl derivative used commercially. Because of its hazard it is not available to
the homeowner.   ETHYL PARATHION

$$(C_2H_5O)_2\overset{\overset{S}{\|}}{P}-O-\langle\ \rangle-NO_2$$

0,0-diethyl 0-p-nitrophenyl phosphorothioate

    Methyl parathion became available in 1949, and proved
to be more useful than (ethyl) parathion because of its
lower toxicity to man and domestic animals and broader
range of insect control. Its shorter residual life also
makes it more desirable in certain instances. This insecticide is also not used by the lay person.

                    METHYL PARATHION

$$(CH_3O)_2\overset{\overset{S}{\|}}{P}-O-\langle\ \rangle-NO_2$$

0,0-dimethyl 0-p-nitrophenyl phosphorothioate

    Closely related to methyl parathion is fenitrothion,
(structure not shown) which is marketed by three basic manufacturers as Accothion[R], Folithion[R], and Sumithion[R]. Controls chewing and sucking insects on fruits, vegetables,
cereals, cotton, rice, and is used in public health vector
control programs. Its use in the U. S. is limited.

    Still another phenyl derivative closely related to
methyl parathion is fenthion (structure not shown) marketed
in the U. S. under several names, each for different purposes. Baytex[R] for ornamentals, mosquito and fly control;
Entex[R] for pest control operator use; and Tiguvon[R] as a non-
systemic pour-on for cattle grubs, lice and flies. Fenthion
has been used very effectively, but illegally, for control
of grain scavenging birds around grain storage and cattle
feed lots by painting the upper surfaces of railings and
other perches with the concentrate.

    Systemic insecticides are also found in the phenyl OPs.
Two examples not illustrated, ronnel and crufomate, were
animal systemics used for the control of cattle grubs. Unfortunately, the manufacture of both materials has been discontinued.

Famphur is an effective systemic insecticide that is applied dermally to livestock as a pour-on formulation. It controls cattle grubs and lice, and is one of the few remaining systemics useful against the cattle grub.

FAMPHUR (Warbex[R])

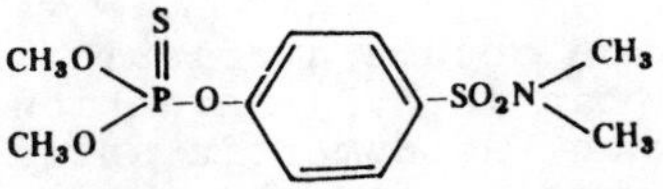

CH_3O

CH_3O

S

$P-O-$

$-SO_2N$

CH_3

CH_3

0,0-dimethyl O-(para-(dimethylsulfamoyl)
phenyl) phosphorothioate

Though not considered a systemic, coumaphos (not shown) is marketed under several names, Coral[R], Meldane[R], and Baymix[R]. It is registered for control of cattle grub and other pests on livestock, and for worm control in cattle and chickens as a feed additive.

Tetrachlorvinphos is useful for fruit and vegetable insects, gypsy moth on forest and shade trees, for area control of ticks, and pests of stored agricultural products.

5.30 What is the difference between the mode of action of the aliphatic and phenyl OP derivatives?

Sulprofos and profenofos are two of the more recently registered phenyl derivatives. Both materials have a broad spectrum of insecticidal activity and are currently labeled for use only on field crops. Isofenphos is used as a soil insecticide in field crops and vegetables, against corn rootworm and onion maggot, and also for white grubs, chinch bugs and sod webworms in turf.

Bolstar^R and Curacron^R are two of the more recently registered phenyl derivatives. Both materials have a broad spectrum of insecticidal activity and are currently labeled for use only on field crops. Isofenphos is used as a soil insecticide in field crops and vegetables, against corn rootworm and onion maggot, and also for white grubs, chinch bugs and sod webworms in turf.

TETRACHLORVINPHOS (Gardona^R, Rabon^R)

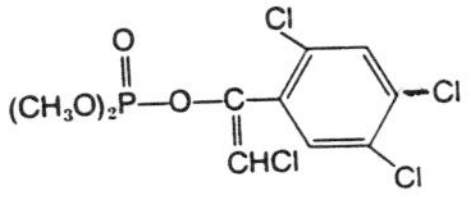

0,0-dimethyl 0-2-chloro-
1-(2,4,5-trichlorophenyl)
vinyl phosphate

SULPROFOS (Bolstar^R)

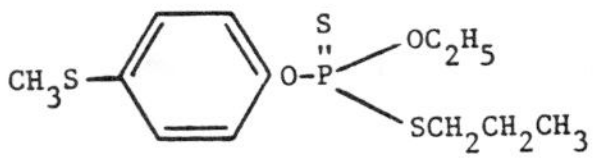

0-ethyl S-propyl 0(4-methylthio)
phenyl phosphorodithioate

PROFENOFOS (Curacron^R)

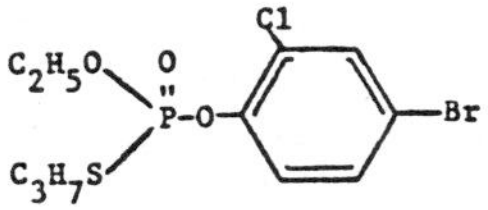

0-(4-bromo-2-chlorophenyl)-0-ethyl
S-propyl phosphorothioate

ISOFENPHOS (Amaze^R)

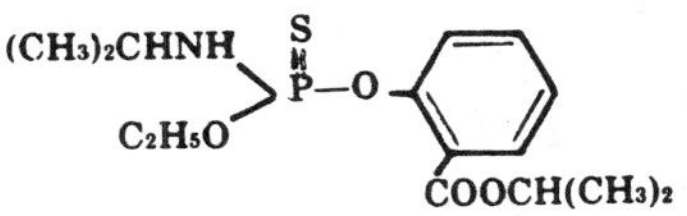

1-methylethyl 2-((ethoxy((1-methyl-
ethyl) amino) phosphinothioyl) oxy)benzoate

Heterocyclic derivatives. The term *heterocyclic* indicates that the ring structures are composed of *unlike atoms*. In a heterocyclic carbon compound, for example, one or more of the carbon atoms is displaced by oxygen, nitrogen or sulfur, and the ring may have 3, 5, or 6 atoms.

5.31 Heterocyclic indicates that cyclic molecules are made
     of _____ atoms.

- -

     different or unlike

Diazinon is probably the first (1952) insecticide made available in this group. Note that the 6-membered ring contains 2 nitrogen atoms, very like the source of its proprietary name, since one of the constituents used in its manufacture is pyrimidine, a diazine. It is registered for virtually all crops and is used extensively as a home and garden insecticide.

DIAZINON (Spectracide[R], Knox Out[R])

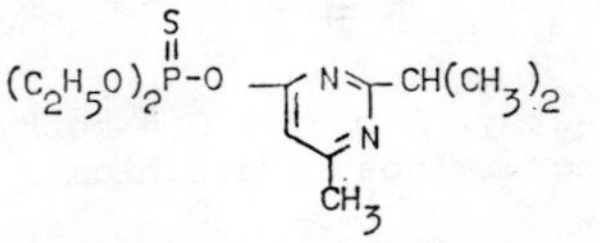

0,0-diethyl 0-(2-isopropyl-4-methyl-6-pyrimidyl) phosphorothioate

Azinphosmethyl is the second oldest member of this group (1954) used in U. S. agriculture. It serves both as an insecticide and acaricide on cotton, vegetables, fruit, nut, field crops, ornamentals, and forest and shade trees.

AZINPHOSMETHYL (Guthion[R])

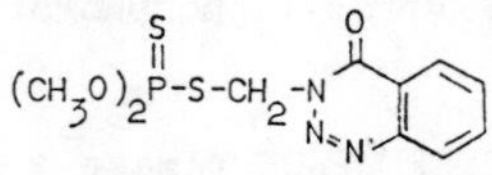

0,0-dimethyl S(4-oxo-1,2,3-benzotriazin-

3(4H)-ylmethyl) phosphorodithioate

Chlorpyrifos is the most frequently used insecticide by pest control operators in homes and restaurants for cockroach and other household insect control, turf and ornamental pests. It is now registered for subterranean termite control, and is used in agriculture as a soil insecticide for corn, and foliar application on cotton, soybeans, and other field crops, as well as for fruit, nuts and vegetables.

CHLORPYRIFOS (Dursban[R], Lorsban[R])

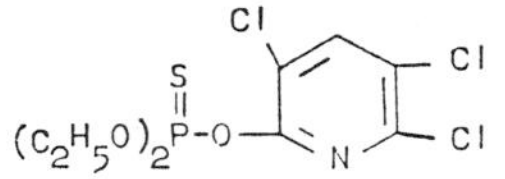

O,O-diethyl O-(3,5,6-trichloro-2-pyridyl) phosphorothioate

Introduced in the mid-1960s methidathion has in the last few years acquired registrations for forage and field crops, tree fruits, and nut crops for a wide range of insect and mite pests. Phosmet is registered for control of the infamous bollweevil and the plum curculio, two closely related weevil pests.

METHIDATHION (Supracide[R])

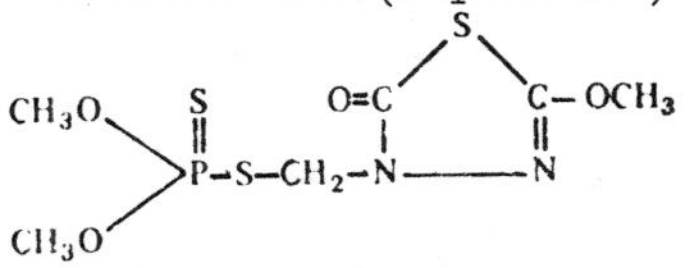

O,O-dimethyl phosphorodithioate,S-ester with 4-(mercaptomethyl)-2-methoxy-Δ^2-1,3,4-thiadiazolin-5-one

PHOSMET (Imidan[R])

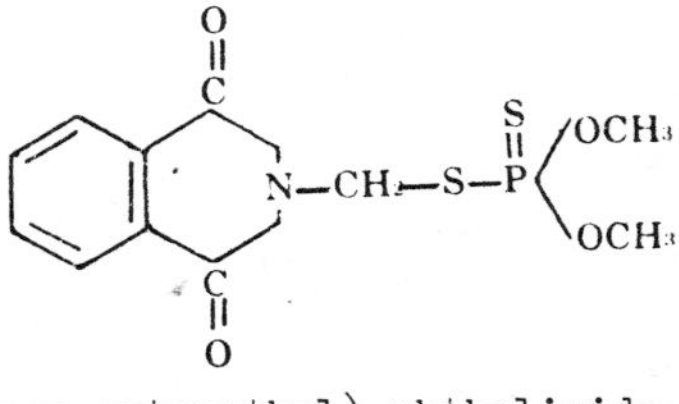

N-(mercaptomethyl)-phthalimide S-
(O,O-dimethyl phosphorodithioate)

The heterocyclic organophosphates are complex molecules, and generally have longer lasting residues than many
of the aliphatic or phenyl derivatives. Also, because of
the complexity of their molecular structures, their breakdown products (metabolites) are frequently many, making
their residues sometimes difficult to measure in the laboratory. Consequently, their use on food crops is somewhat
less than either of the other two classes of OPs.

5.32 Using the three structures, identify the OP category:
     aliphatic_____   phenyl_____   heterocyclic _____

(1)   (2)   (3)

$(C_2H_5O)_2\overset{S}{\underset{\|}{P}}-O-N=\overset{CN}{\underset{|}{C}}$ —phenyl

$(C_2H_5O)_2\overset{S}{\underset{\|}{P}}-O$ —quinoxaline (2)

$(C_2H_5O)_2\overset{O}{\underset{\|}{P}}-S-CH_2\overset{O}{\underset{\|}{C}}-NH-\overset{CN}{\underset{|}{C}}(CH_3)_2$

- -

     aliphatic 3   phenyl 1   heterocyclic 2

5-C ORGANOSULFURS

As the name suggests, the organosulfurs have sulfur
as their central atom. They resemble the DDT structures
in that most have two phenyl rings.

5.34 The organosulfurs have _____ as the
     central atom on which the remainder of the molecule
     is attached.

- -

     sulfur

Sulfur alone is a good acaricide (miticide), but it is
also phytotoxic, (particularly in hot weather), and may be
used at the rate of 10 to 30 lb per acre for this purpose.
The organosulfurs are far superior and require only 1 to 2
lb per acre. You may have observed that sulfur in combination with phenyl rings is particularly toxic to mites and
your observation is accurate. Of greater interest, however,
is that the organosulfurs have very low toxicity to insects.
As a result they are used for selective mite control in
integrated pest management systems.

5.35 The organosulfurs are used almost exclusively as

_____.

- -

   miticides or acaricides

    This group of miticides has one other valuable quality:
they are usually ovicidal as well as being toxic to the
young and adult mites.

5.36 An ovicide, then, must be a material that kills

_____.

- -

   eggs

    Tetradifon is one of the older acaricides and bears
typically the sulfur and twin phenyl rings, as do most of
the organosulfurs.

TETRADIFON (Tedion®)

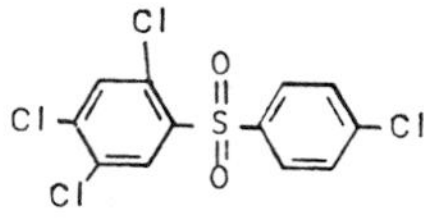

p-chlorophenyl 2,4,5-trichlorophenyl sulfone

PROPARGITE (Omite^R)

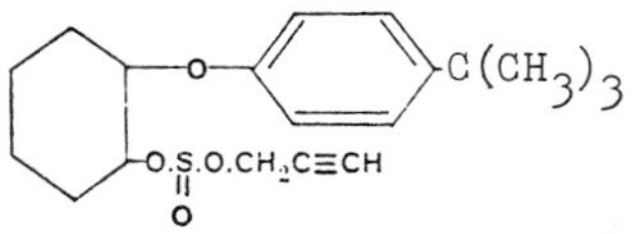

2-(p-tert-butyl phenoxy)cyclohexyl 2-propynyl sulfite

OVEX (Ovotran^R)

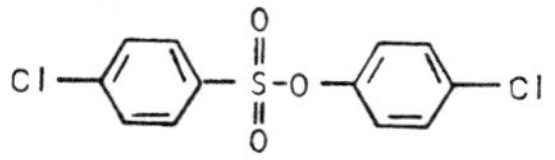

p-chlorophenyl p-chlorobenzenesulfonate

5-D   CARBAMATES

If the phosphate insecticides are derivatives of
phosphoric acid, then the carbamates must be derivatives
of carbamic acid $\overset{O}{\underset{}{\overset{\|}{C}}}$. And, like the organophosphates,

$$HO-\overset{O}{\overset{\|}{C}}-NH_2$$

the mode of action of the carbamates is that of inhibiting
that vital enzyme, cholinesterase (ChE).

5.37 The carbamate insecticides are esters of _____
     acid.

- -

     carbamic

5.38 Carbamates kill by inhibiting _____.

- -

     ChE

5.39 Of the three major groups of insecticides, organo-
     chlorines, organophosphates, and carbamates, which
     two have generally similar modes of action?

     _____
     _____

- -

     organophosphates and carbamates
     (inhibition of ChE)

The carbamate insecticides were introduced in 1951 by
the Geigy Chemical Company of Switzerland but fell by the
wayside because the first were ineffective. Two of the
early carbamates are presented below.

ISOLAN   DIMETAN

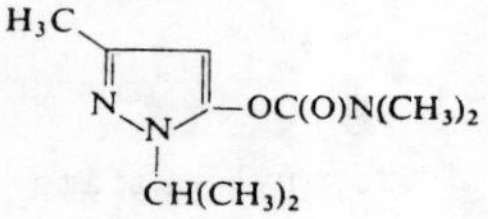

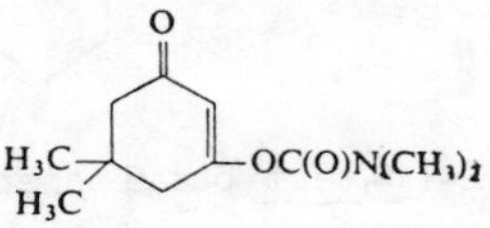

H_3C O

N-N—OC(O)N(CH$_3$)$_2$ H$_3$C— —OC(O)N(CH$_3$)$_2$

CH(CH$_3$)$_2$ H$_3$C

It was not known at that time that the N,N-dimethyl
carbamates, as above, were generally less toxic to insects
than the N-methyl carbamates, which were developed later
and which are the bulk of the currently used materials.

The first successful carbamate was carbaryl, introduced in 1956. More of it has been used world over than all the remaining carbamates combines. Two distinct qualities have made it the most popular material: very low mammalian oral and dermal toxicity, and a rather broad spectrum of insect control. This has led to its wide use as a lawn and garden insecticide.

CARBARYL (Sevin®)

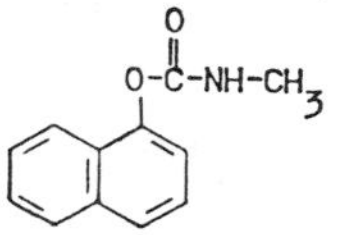

$$O-\overset{\overset{\displaystyle O}{\|}}{C}-NH-CH_3$$

1-naphthyl methylcarbamate

Methomyl is a more recent carbamate that has been extremely useful, especially for worm control on vegetables. Much of this efficacy can be attributed to its ovicidal properties (kills the eggs). Closely related is oxamyl.

METHOMYL (Lannate[R], Nudrin[R])

$$CH_3-\underset{\underset{\displaystyle S-CH_3}{|}}{C}=N-O-\overset{\overset{\displaystyle O}{\|}}{C}-NH-CH_3$$

methyl N-((methylcarbamoyl)oxy)thioacetimidate

OXAMYL (Vydate[R])

$$(CH_3)_2N\overset{\overset{\displaystyle O}{\|}}{C}-\underset{\underset{\displaystyle S-CH_3}{|}}{C}=N-O-\overset{\overset{\displaystyle O}{\|}}{C}-NH-CH_3$$

methyl N',N'-dimethyl-N-((methylcarbamoyl)-
oxy)-1-thiooxamimidate

Several of the carbamates are plant systemics, indicating that they have a rather high water solubility in order to be taken into the roots or leaves. They are also not readily metabolized by the plants. Methomyl, aldicarb and carbofuran have distinct systemic characteristics, making them useful also as nematicides. Of these, only aldicarb and carbofuran are used as soil insecticides/nematicides. Aldicarb has received much attention in the mid-'80s due to its chemical stability in some soils, allowing it to appear chemically intact in shallow ground waters.

ALDICARB (Temik®)

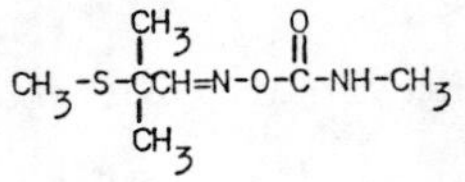

$$CH_3-S-\overset{\overset{\displaystyle CH_3}{|}}{\underset{\underset{\displaystyle CH_3}{|}}{C}}CH=N-O-\overset{\overset{\displaystyle O}{||}}{C}-NH-CH_3$$

2-methyl-2-(methylthio) propionaldehyde O-(methylcarbomoyl) oxime

CARBOFURAN (Furadan®)

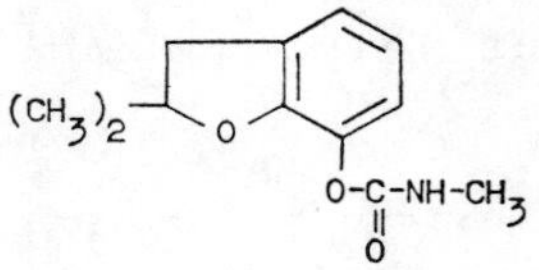

$$O-\overset{}{\underset{\underset{\displaystyle O}{||}}{C}}-NH-CH_3$$

2,3-dihydro-2,2-dimethyl-7-benzofuranyl methylcarbamate

5.40 To be a plant systemic a pesticide must have two basic characteristics. They have a high _____ solubility, and are not readily _____ by the plant.

- -

water   metabolized

Thiodicarb (Larvin[R]) is the most recently registered carbamate for agricultural use. Approval was granted by EPA in 1985 for use on cotton, soybeans, field corn and processing sweet corn. Effective against essentially all types of caterpillars, it possesses an extremely enduring residue, often up to 14 days. Thiodicarb is really a "double carbamate", two carbamate molecules linked with a sulfur atom. Because of its novel structure and long residual, it has been referred to as a "second generation carbamate insecticide".

THIODICARB (Larvin^R)

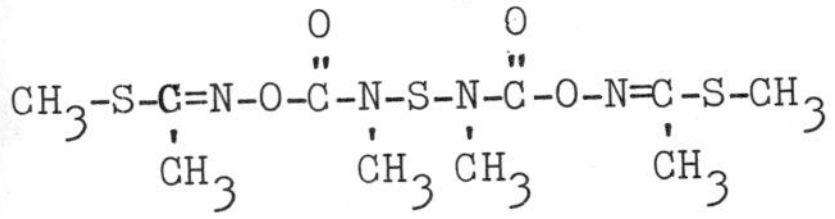

$$CH_3-S-\underset{\underset{CH_3}{|}}{C}=N-O-\overset{\overset{O}{||}}{C}-\underset{\underset{CH_3}{|}}{N}-S-\underset{\underset{CH_3}{|}}{N}-\overset{\overset{O}{||}}{C}-O-N=\underset{\underset{CH_3}{|}}{C}-S-CH_3$$

dimethyl N N' thiobis(methylimino)
carbonyloxy bis ethanamidothioate

Another carbamate, propuxur, is highly effective
against cockroaches that have developed resistance to the
organochlorines and organophosphates. Propoxur is used by
structural pest control operators for roaches and most
other household insects in restaurants, kitchens, and homes.
Similarly, bendiocarb has found its greatest use as a residual household insecticide (Ficam^R) by pest control applicators, and for turf and ornamental pests (Turcam^R).

PROPOXUR (Baygon$^®$)

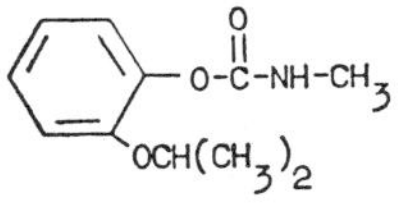

o-isopropoxyphenyl methylcarbamate

BENDIOCARB (Ficam^R, Turcam^R)

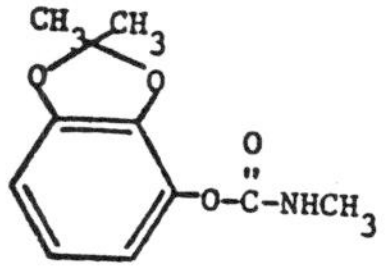

2,2-dimethyl-1,3-benzodioxol-4-yl methylcarbamate

Bufencarb is used in agriculture exclusively as a soil
insecticide, becoming a replacement for the long-residual
organochlorine insecticides, aldrin, dieldrin and heptachlor. Methiocarb, aminocarb and promecarb are effective
against foliage and fruit eating insects. Methiocarb and
aminocarb are both excellent molluscicides, used for slug
and snail control in flower gardens and ornamentals.
Methiocarb has the additional distinction of being registered as a bird repellent for cherries and blue berries and
as a seed dressing.

BUFENCARB (Bux^R)

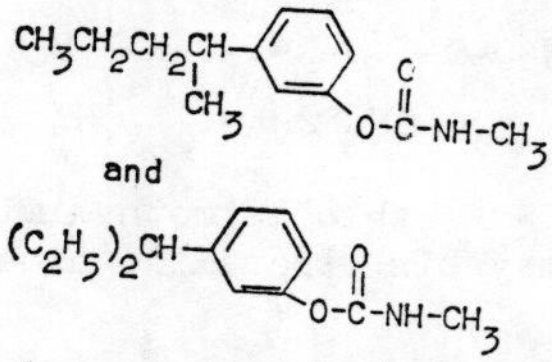

and

mixture of m-(ethylpropyl)phenyl methylcarbamate
and m-(1-methylbutyl)phenyl methylcarbamate (ratio 1:3)

METHIOCARB (Mesurol^R)

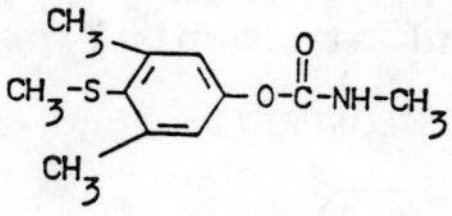

4-(methylthio)3,5-xylyl methylcarbamate

AMINOCARB (Matacil^R)

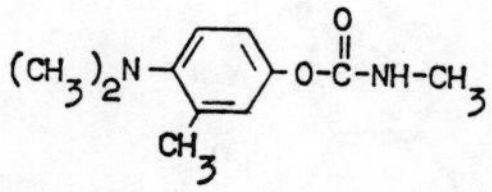

4-(dimethylamino)-3-methylphenol methylcarbamate

PROMECARB (Carbamult^R)

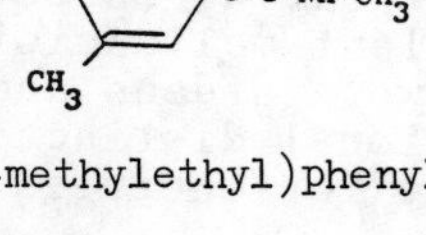

3-methyl-5-(1-methylethyl)phenyl methylcarbamate

In summary, the carbamates are inhibitors of cholinesterase, are plant systemics in several instances, and
are, for the most part, broad-spectrum in effectiveness,
being used as insecticides, miticides, molluscicides, and
nematicides.

5-E FORMAMIDINES

Little will be said about this small but promising
group of insecticides. Three examples are chlordimeform,
formetanate, and amitraz. They are effective against the
eggs and very young caterpillars of several moths of agricultural importance and are also effective against most
stages of mites and ticks. Thus, they are classed as
ovicides, insecticides, and acaricides. Late in 1976,
chlordimeform was removed from the market by its manufacturers, CIBA-GEIGY Corporation, and Nor-Am Agricultural
Products, Inc., because it proved to be carcinogenic to a
cancer-prone strain of laboratory mice during high-level,
lifetime feeding studies. In 1978 it was returned for use
on cotton, but under strict application restrictions.

CHLORDIMEFORM (Galecron®, Fundal®)

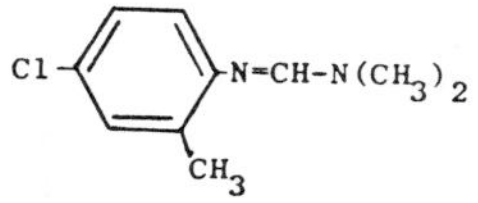

N'-(4-chloro-o-tolyl)-N,N-dimethylformamidine

AMITRAZ (Baam^R)

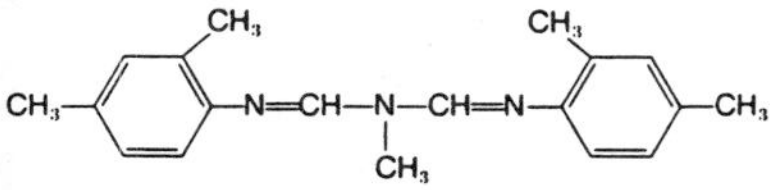

N'-(2,4-dimethylphenyl)-N-(((2,4-dimethyl=
phenyl)imino)methyl)-N-methylmethanimidamide

FORMETANATE (Carzol©)

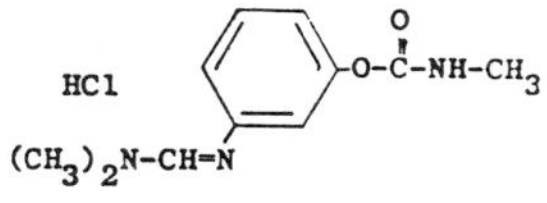

(3-dimethylamino-(methylene-iminophenyl))
-N-Methylcarbamate hydrochloride

Their present value lies in the control of organophos-
phate- and carbamate-resistant pests. Poisoning symptoms
are distinctly different from other materials. It has been
currently proposed that one possible mode of action is the
inhibition of a previously unmentioned enzyme, *monoamine
oxidase*. This results in the accumulation of compounds
termed *biogenic amines*. Thus, the formamidines introduce
a new mode of action for the insecticides and acaricides.
This makes them extremely useful, for we are slowly losing
ground in the battle of insect resistance to the modes of
action of the older insecticide groups.

5.41 A new mode of action for the insecticides is revealed
      for the formamidines, the inhibition of _____
      _____.

----

monoamine oxidase

5-F AMIDINOHYDRAZONES

There is only one member of this group for the moment,
pyramdron. However, with its success as an ant and cockroach
bait there surely will be more. It is a stomach poison
marketed under the names Amdro[R], for imported fire ants,
harvester ants and big-headed ants, and Maxforce[R] for control of all pest species of cockroaches.

PYRAMDRON (Amdro[R], Maxforce[R])

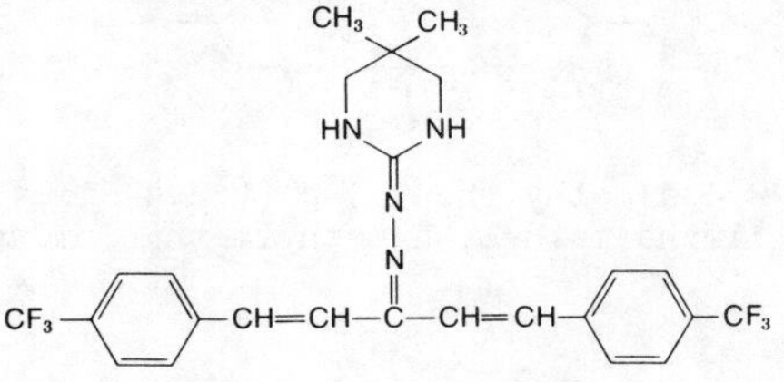

tetrahydro-5,5-dimethyl-2(1H)-
pyrimidinoine (3-|4-(trifluoromethyl)-
phenyl|-1-(2-|4-(trifluoromethyl)phenyl|-
ethenyl)-2-propenylidene) hydrazone

5-G   DINITROPHENOLS

The dinitrophenols are another group possessing easily recognized structural formulas. Di (two) nitro (NO_2)
phenol [structure] , or easier yet, dinitrophenol: [structure]
The basic dinitrophenol molecule has a broad range of
toxicities. Compounds derived from it are used as herbicides, insecticides, ovicides, and fungicides. They act
by *uncoupling oxidative phosphorylation,* or basically
preventing the utilization of nutritional energy. In the
1930s, certain dinitrophenols were given by uninformed
physicians to their overweight patients to induce rapid
weight loss. They were extremely effective, but quite
toxic, and their use resulted in several widely publicized
deaths.

5.42 Besides insecticides, what other pesticidal uses are
     made of dinitrophenols?

----

     herbicides, fungicides, ovicides

5.43 The dinitrophenol mode of action prevents utilization
     of energy by blocking _____ _____.
----

     oxidative phosphorylation

The oldest of this group is DNOC (3,5-dinitro-0-
cresol), introduced as an insecticide in 1892. DNOC has
also been used as an ovicide, herbicide, fungicide, and
blossom thinning agent. Its use has declined today to
herbicidal applications where all plants are to be killed.
Dinoseb is used as a dormant fruit spray for control of
many insects and mites.

DINITROCRESOL (DNOC)   DINOSEB

[structure]   [structure]

4,6-dinitro-o-cresol   2-sec-butyl-4,6-dinitrophenol

Binapacryl, or Morocide[R], is used exclusively as an acaricide and was introduced in 1960. Dinocap was developed in 1949 as an acaricide and fungicide, and is one of the rare materials made up of several related molecular structures, only one of which is shown. Dinocap is particularly effective against powdery mildew fungi. Owing to its safety to green plants, it has often replaced the phytotoxic sulfur which is so effective against powdery mildews.

BINAPACRYL (Morocide[®])

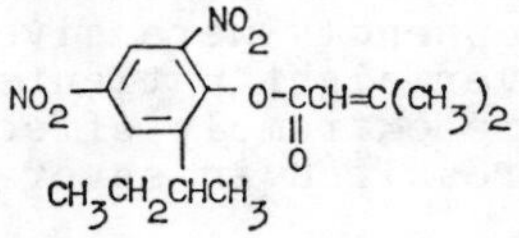

2-sec-butyl-4,6-dinitrophenyl 3-methyl-2-butenoate

DINOCAP (Karathane[®])

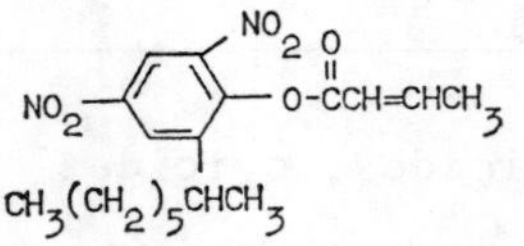

2-(1-methylheptyl)-4,6-dinitrophenyl crotonate

In summary, the dinitrophenols have been used in practically all pesticide classifications: ovicides, insecticides, acaricide, herbicides, fungicides, and blossom thinning agents.

5-H   ORGANOTINS

The sole purpose of this short entry is to introduce a relatively new group of acaricides, which double as fungicides, as you will see later. Of particular interest here is Plictran[R], one of the most selective acaricides presently known, introduced in 1967. Introduced somewhat later, Vendex[R] has proved to be most effective against mites on deciduous fruits, citrus, greenhouse crops and ornamentals. The mode of action of this group is not completely known, but is believed to be the inhibition of oxidative phosphorylation at the site of dinitrophenol uncoupling. This inhibition reduces substantially, and in the case of mites, fatally, the availability of energy in the form of adenosine triphosphate (ATP). These trialkyl tins also inhibit photophosphorylation in chloroplasts, and can thus serve as algicides.

CYHEXATIN (Plictran[R])

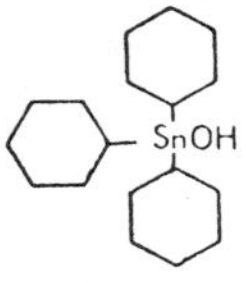

tricyclohexylhydroxytin

FENBUTATIN-OXIDE (Vendex[R])

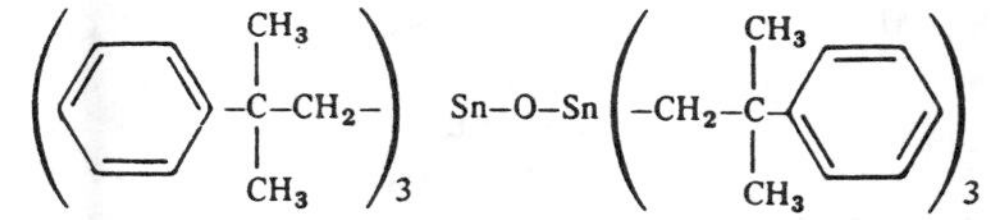

hexakis (2-methyl-2-phenylpropyl)distannoxane

5.44 The organotins' mode of action is probably the inhi-
     bition of _____ _____.
- -
     oxidative phosphorylation

5-I BOTANICALS

The botanicals or "natural" insecticides are toxicants derived from plants. Historically, the plant materials have been in use longer than any other group, with
the possible exception of sulfur. Tobacco, pyrethrum,
derris, hellebore, quassia, camphor, and turpentines were
some of the more important plant products in use before
the organized search for insecticides had begun.

5.45 Organic gardeners who use insecticides normally
     select the botanicals. Why?

- -
     plant-derived and considered of "natural" origin

Some of the most universally used insecticides are
derived from plants. The flowers, leaves, or roots have
been finely ground and used in this form, or the toxic
ingredients have been extracted and used alone or in mixtures with other toxicants. Of the botanicals, only

nicotine, pyrethrum, rotenone, sabidilla, and ryania will
be discussed in this manual.

Sir Walter Raleigh introduced smoking tobacco to
England in 1585. As early as 1690 water extracts of tobacco were reported as being used to kill sucking insects
on garden plants. As early as about 1890, the active principle in tobacco extracts was known to be nicotine, and,
from that time on, extracts were sold as commercial insecticides for home, farm, and orchard. Today organic gardeners may soak a cigar or two in water overnight and spray
insect-infested plants with the extract, achieving some
success. "Black Leaf 40," which has long been a favorite
garden spray, is a concentrate containing 40% nicotine
sulfate. Today, nicotine is commercially extracted from
tobacco by steam distillation or solvent extraction.

NICOTINE

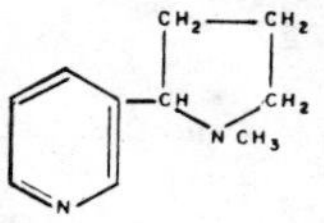

1-3-(1-methyl-2-pyrrolidyl) pyridine

Nicotine is an *alkaloid*; it is a heterocyclic compound containing nitrogen and having prominent physiological properties. Other well-known alkaloids, which
are not insecticides, are caffeine (found in tea and cof-
fee), quinine (from cinchona bark), morphine (from the
opium poppy), cocaine (from coca leaves), ricinine (a
poison in castor oil beans), strychnine (from *Strychnos
nux-vomica*), coniine (from spotted hemlock, the poison
that killed Socrates), and, finally, LSD (from the ergot
fungus attacking grain), one of the banes of our 20th-
century culture.

5.46 Alkaloids are heterocyclic compounds containing
_____ and are _____
active.

- -

nitrogen   physiologically

5.47 Several other well-known alkaloids are _____,
_____, _____ and _____.

- -

caffeine, quinine, morphine, cocaine, strychnine

Nicotine mimics acetylcholine (ACh) at the neuromuscular (nerve-muscle) junction in mammals, and results in
twitching, convulsions, and death, all in rapid order.   In
insects, the same action is observed, but only in the
ganglia of their central nervous systems.

5.48 Nicotine mimics _____ as its
     mode of action.
- -

     acetylcholine (ACh)

Rotenone or the rotenoids have been used as crop insecticides since 1848, when they were applied to plants
to control leaf-eating caterpillars.   However, they have
been used for centuries (at least since 1649) in South
America to paralyze fish causing them to surface.
Rotenoids are produced in the roots of two genera
of the legume (bean) family, *Derris*, grown in Malaya and
the East Indies, and *Lonchocarpus* (also called *cubé*),
grown in South America.

5.49 Rotenone is the principal toxic compound in _____
     and _____.
- -

     derris   cubé   Lonchocarpus

Rotenone is a very safe garden insecticide, resulting
in its popularity for over the last 60 years.   It is highly
toxic to most insects with chewing mouthparts, and its mode
of action is the interference in energy production.   It is
a respiratory enzyme inhibitor, acting between NAD^+ and
coenzyme Q, resulting in failure of the respiratory functions.   Its extreme toxicity to fish and insects and its
low toxicity to mammals is explained by the fact that
insects and fish convert rotenone to highly toxic metabolites, while mammals do not.

ROTENONE

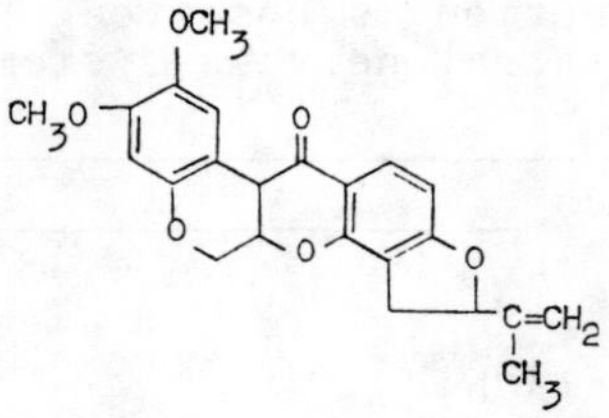

1,2,12,12a,tetrahydro-2-isopropenyl-8,9-dimethoxy-[1]

benzopyrano-[3,4-b]furo[2,3-b][1]benzopyran-6(6aH)one

5.50 Rotenone exerts its action by blocking _____

_____.

- -

oxidative phosphorylation

Rotenone is now used almost exclusively for reclaiming
lakes for game fishing. It eliminates all fish, closing
the lake to reintroduction of rough species. After treatment the lake can be restocked with the desired species.
Rotenone is a selective piscicide in that it kills all fish
at dosages that are relatively non-toxic to fish food organisms. It also breaks down quickly leaving no residues
harmful to the fish used for restocking. The recommended
rate is 0.5 parts of rotenone to one million parts of
water (ppm), or 1.36 pounds per acre-foot of water.

5.51 Two good reasons why rotenone is an outstanding
piscicide:

_____

_____

- -

1. kills fish but not fish food
2. degrades rapidly

Ryania was introduced in 1945, and made by grinding
the stem of a tropical shrub, *Ryania speciosa,* native to
Trinidad and the Amazon basin. It was a useful insecticide in its time, but was displaced by the less expensive synthetic organochlorines. The active principal is
the alkaloid ryanodine, which affects muscles directly

by preventing contraction, thereby resulting in paralysis,
resembling the effects of strychnine in mammals. Ryanodine
is at least twenty times more toxic to mammals than to most
insects. This certainly contradicts the impression that
natural insecticides are in some way safer to use than
synthetic compounds. The mode of action of ryanodine is
that of membrane disruption rather than inhibiting enzymes.
Insects increase oxygen consumption followed by flaccid
paralysis and death. In mammals there is progressive
rigidity of the muscles with final respiratory failure. The
structure of ryanodine is not known. Ryania is no longer
available in the U. S.

RYANIA

ryanodine $C_{25}H_{35}NO_9$

5.52 Ryanodine is an _____ and affects
_____ tissue by _____.

- -

alkaloid   muscle   membrane disruption

Sabadilla is an alkaloid mixture of cevadine and
veratridine, taken from the seeds of *Schoenocaulon of-
ficinale*, a member of the lily family grown in Venezuela.
Sabadilla has been used variously in the home and garden,
and is known for its absolute safety when used around
warm-blooded animals. There is one undesirable characteristic — it is an irritant to mucous membranes and causes
violent sneezing in some individuals. Sabadilla too,
probably acts on muscle tissue, resulting in paralysis.
Purified cevadine is approximately ten times more toxic
than DDT to houseflies. Veratridine produces a prolonged
rigor in mammalian skeletal muscle. Oxygen consumption
increases, but not to the extent noted in ryanodine poisoning. The mode of action for sabadilla in insects appears
to be the same as ryanodine, resulting in flacid paralysis
and death. The structures for sabadilla are also not
known. Sabadilla is no longer available in the U. S.

SABADILLA

cevadine $C_{32}H_{49}NO_9$

veratridine $C_{36}H_{51}NO_{11}$

5.53 Sabadilla is an _____ belonging to the
     botanical insecticides.

- -

     alkaloid

     Pyrethrum is extracted from flower heads of Chry-
santhemum in the daisy family. Pyrethrum powder was first
used as an insecticide in the Transcaucasus region of Asia
about 1800. Originally, the flower heads were ground and
used in that form for louse control in the Napoleonic Wars.
Kenya, Africa, is now the primary source of pyrethrum,
which is now extracted from the flower heads with solvents.
     Pyrethrum sprays are ideal home insecticides because
of their extremely rapid knockdown (the instant action much
favored by the American public) and they are probably the
safest insecticides known for man and his domestic animals.
When pyrethrum is used alone, downed insects may recover
and return to strike again. Consequently, synergists were
discovered for the sole purpose of preventing insect sur-
vivors. The synergist prevents the insect from metabolizing pyrethrum and recovering.

5.54 Pyrethrum is perfect for home use. It offers rapid
     _____ and is the _____
     insecticide known for use in the home environment.

- -

     knockdown   safest

5.55 Synergists prevent insects from _____
     pyrethrum.

- -

     metabolizing

     Pyrethrum is a mixture of four compounds: pyrethrins
I and II and cinerin I and II. Their structures can be
assembled by attaching the R_1 and R_2 in their proper posi-
tions on the large ester structure to the left.

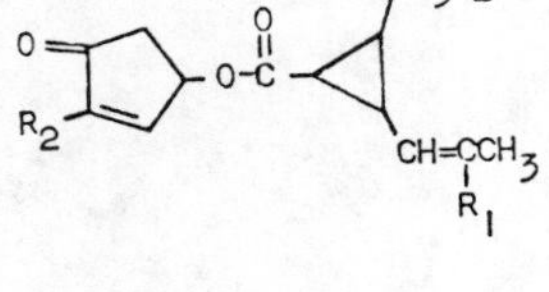

Pyrethrin I:
R_1 = -CH$_3$
R_2 = -CH$_2$CH=CHCH=CH$_2$

Pyrethrin II:
R_1 = -C-OCH$_3$ (=O)
R_2 = -CH$_2$CH=CHCH=CH$_2$

Cinerin I:
R_1 = -CH$_3$
R_2 = -CH$_2$CH=CHCH$_3$

Cinerin II:
R_1 = -C-OCH$_3$ (O)
R_2 = -CH$_2$CH=CHCH$_3$

Pyrethrum is considered an axonic poison, as are all or the organochlorine insecticides. The axon of a nerve cell is a long extension of the cell body and is vital in the transmission of nerve impulses from one cell body to other cells. Chemicals that affect this impulse transmission are referred to as axonic. Pyrethrum has a greater insecticidal effect when the temperature is lowered, referred to as a negative temperature coefficient. The fast knockdown of flying insects is the result of rapid muscular paralysis, making it appear to have its effect on the ganglia of the insect central nervous system. There is also evidence that its effects are on the neurons.

5.56 Pyrethrum has a _____ temperature coefficient.

----

negative

The use of botanical insecticides reached its maximum in the U. S. in 1966, and has declined steadily since. Pyrethrum is now the only botanical of significance in use, and this is typically in rapid knockdown sprays in combination with synergists and one or more synthetic organic insecticides which are formulated for use in the home and garden.

In summary, these "natural" insecticides are chemicals and really no safer than most of the currently available synthetic insecticides, at least as compared to those available to the layman. Their only distinction is that they are synthesized by plants.

5.57 List the 5 useful botanicals just studied.

----

nicotine   pyrethrum   rotenone   sabadilla   ryania

5-J SYNTHETIC PYRETHROIDS

Pyrethrum was almost never used for agricultural purposes because of its cost and instability in sunlight. Recently, however, several synthetic pyrethrin-like materials have become available only to agriculture and are referred to as synthetic pyrethroids. These materials are very stable in sunlight and are generally effective against most agricultural pests when used at the low rates of 0.1 to 0.2 pound per acre. Examples are permethrin (Ambush[R], or Pounce[R]) and fenvalerate (Pydrin[R]).

PERMETHRIN (Ambush[R], Pounce[R])

m-phenoxybenzyl(±)-cis,trans-3-(2,2-dichlorovinyl)-
2,2-dimethylcyclopropanecarboxylate

FENVALERATE (Pydrin[R])

cyano (3-phenoxyphenyl)methyl 4-chloro-
a-(1-methylethyl) benzeneacetate

The pyrethrin-like compounds have a long and successful history. I have placed them in clusters or "generations" for ease of organization.
Generation 1: This contains only one pyrethroid,
allethrin, which appeared in 1949. Allethrin was a synthetic duplicate of Cinerin I (see Pyrethrum).
Generation 2: Here we were introduced to tetramethrin
(Neo-Pynamin[R]) in 1965, resmethrin (NRDC-104, SBP-1382, and
FMC-17370) in 1967, bioresmethrin (NRDC-107, FMC-18739, and
RU-11484) in 1967. Bioallethrin[R] was introduced in 1969,
and phenothrin (Sumithrin[R]) in 1973. None of these were
agriculturally successful, because of their overall lack
of efficacy or their sensitivity to sunlight. However,
allethrin, resmethrin, bioresmethrin and Bioallethrin[R] were
somewhat better than pyrethrum in household pest control
and pyrethrum-containing aerosols. Consequently they have
been used in home and pest control operator formulations
for the past 10 to 12 years.
Generation 3: Fenvalerate and permethrin appeared in
1972 and 1973, respectively. These were stable in sunlight
and only slightly volatile, thus the first real agricultural pyrethroids.
Generation 4: The newer pyrethroids require only
one-fifth to one-tenth of the active ingredient per acre
compared to Generation 3 compounds. An efficacious rate of
0.01 to 0.05 pound per acre is truly phenomenal! Among
these are cypermethrin (Ammo[R], Barricade[R], Cymbush[R], and
Ripcord[R]); cyfluthrin (Baythroid[R]); fenpropathrin (Danitol[R]);
flucythrinate (Pay-Off[R]); fluvalinate (Mavrik[R] and Spur[R]);
tralomethrin (Scout[R]), and bifenthrin (Capture[R]). All of
these are photostable, that is, they do not undergo

photolysis (splitting in the presence of ultra-violet light).
And because they have minimal volatility they provide extended residual effectiveness, sometimes as long as 10 days
under optimum conditions.

All of their chemical structures resemble each other
to some extent, as seen in the illustrations. Fenpropathrin, tralomethrin and bifenthrin are not shown.

The pyrethroids all share similar modes of action,
resembling those of the organochlorine insecticides. They
are considered axonic poisons, affecting electrical impulse
transmissions in axons.

Pyrethroids are more effective when the temperature is
lowered; that is, they have a negative temperature coefficient, as does DDT. They affect both the peripheral and
central nervous system of the insect. Pyrethroids at first
stimulate nerve cells to produce repetitive discharges and
eventually cause paralysis. In some instances the pyrethrum
synergists increase the toxicity of the pyrethroids, but
not nearly to the extent that they enhance the natural pyrethroids.

CYPERMETHRIN (Ammo^R, Barricade^R, Cymbush^R, Ripcord^R)

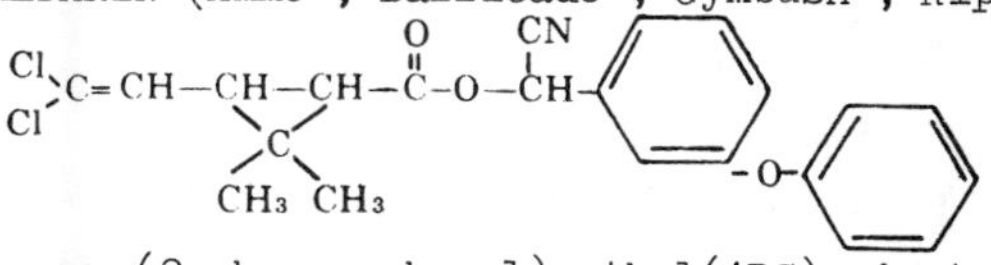

(RS)-cyano(3-phenoxyphenyl)methyl(1RS)-cis-trans-
3-(2,2-dichloroethenyl)-2,2-dimethylcyclopropanecarboxylate

CYFLUTHRIN (Baythroid^R)

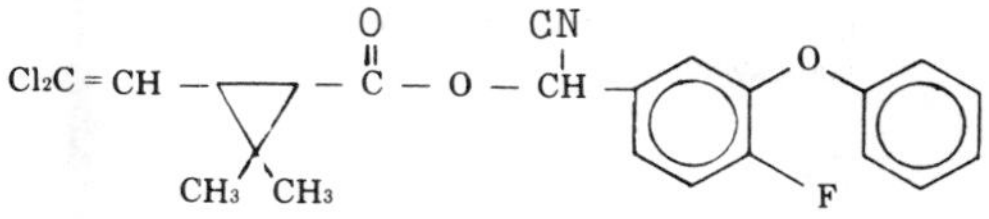

cyano(4-fluoro-3-phenoxyphenyl)methyl-3-(2,2-dichloro-
ethenyl)-2,2-dimethyl-cyclo-propanecarboxylate

FLUCYTHRINATE (Pay-Off[R])

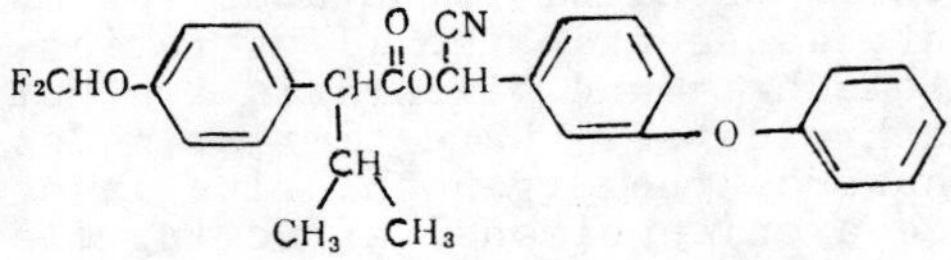

cyano(3-phenoxyphenyl)methyl (+)-4-(difluoromethoxy)-
alpha-(1-methylethyl) benzeneacetate

FLUVALINATE (Mavrik[R], Spur[R])

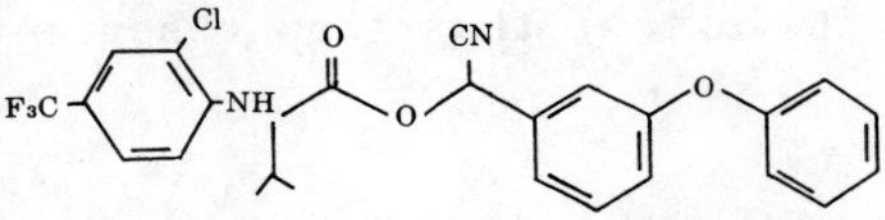

(alpha-RS,2R)-fluvalinate((RS)-alpha-cyano-3-phenoxybenzyl-
(R)-2-(2-chloro-4-(trifluoromethyl)anilino)-3-methyl-butanoate)

BIFENTHRIN (Capture[R], FMC 54800)

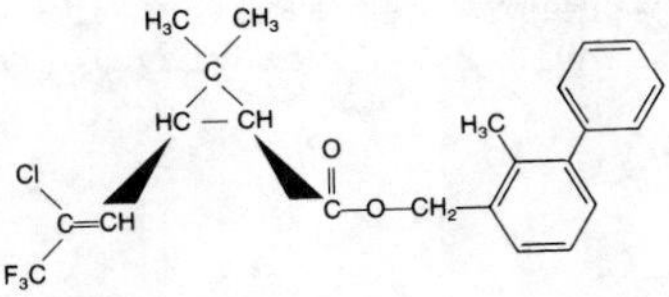

[1α,3α(Z)]-(±)-(2-methyl[1,1'-biphenyl]-3-
yl)methyl 3-(2-chloro-3,3,3-trifluoro-1-pro-
penyl)-2, 2-dimethylcyclopropane-
carboxylate

5.58 The pyrethroids are considered _____
     poisons, and are more effective when the temperature
     is _____.

- -

     axonic   lowered

5-K SYNERGISTS OR ACTIVATORS

     Synergists are considered neither toxic nor insecti-
cidal, but are used with insecticides to enhance or synergize the activity of the insecticides. They are added to
certain insecticides in the ratio of 8:1 or 10:1, syner-
gist:insecticide. The first synergist was introduced in
1940 to increase the effectiveness of pyrethrum. Since
then many materials have been introduced, but only a few
have survived, because of cost and ineffectiveness. Synergists are found in practically all of the "bug-bomb" aerosols to enhance the action of the fast knockdown insecticides pyrethrum, allethrin, and resmethrin against flying
insects.

5.59 A synergist is a chemical added to certain insecti-
     cides to _____ their insecticidal qualities.

- -

     increase or enhance

5.60 Synergists are usually not considered to be _____
     though they increase measurably the toxicity of the
     insecticide.

- -

     toxic

     The synergists were initially developed for use with
pyrethrum, but have since been observed to synergize some,
but not all, organophosphates, organochlorines, carbamates,
as well as a few of the botanicals, or plant-derived insecticides. Several years of mystery surrounded their mode
of action, because to further confuse matters, these compounds produced no effect with some of the insecticide
classes just mentioned, and even antagonized the actions of
others. In other words, synergists could result in a plus,
neutral, or negative effect when added to insecticides. It
has since been well established that the synergists inhibit
certain enzymes involved in the detoxication or in the
activation of insecticides. The mode of action of the

synergists is the inhibition of mixed function oxidases,
enzymes that metabolize foreign compounds, which in this
instance would be the pyrethrum.

5.61 Synergists act by _____ enzymes that nor-
     mally _____ the insecticide.

- -

     inhibiting   detoxify

5.62 For a synergist to antagonize the action of an in-
     secticide, it must _____ an enzyme that
     normally _____ the insecticide.

- -

     inhibit   activates

     The useful synergists, popular in the industry, be-
long to only two molecular groups, or moieties. The first
is the methylenedioxyphenyl moiety. The R_1 and R_2 are
radicals, carbon chains or other groups of
varying combinations, depending on the total
molecular structure. The second synergistic
moiety doesn't have a single name, but is characterized
by either of the following structures.

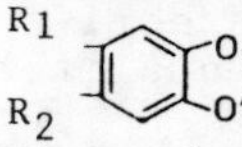

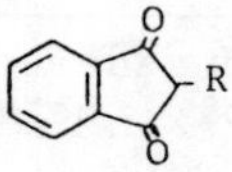

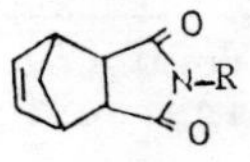

     You have noticed that all three moieties involve a
five-membered ring associated with two oxygens. Because
their mode of action is the inhibition of insecticide-
metabolizing enzymes, it is likely that this steric three-
dimensional structure is generally the most effective in
enzyme binding.
     Synergists are usually used in sprays prepared for the
home and garden, stored grain, and on livestock, particularly in dairy barns. Synergists do not synergize the current group of agricultural pyrethroids to a significant
extent, thus they are not used with them.

5.63 Synergists are not used on agricultural crops. Why?
     _____

- -

     They do not synergize the current pyrethroids

First discovered in sesame oil, a material containing the methylenedioxyphenyl group was given the name sesamin. As mentioned earlier, many compounds having this moiety are synergistic, but only the structures of piperonyl butoxide, Sesamex, and Sulfoxide are shown.

PIPERONYL BUTOXIDE

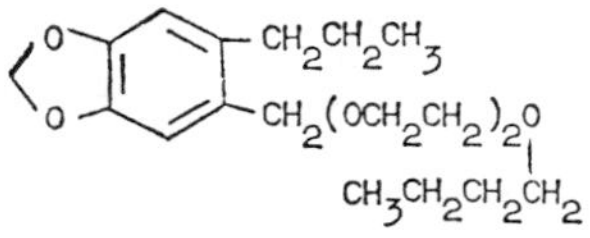

α-[2-(2-butoxyethoxy)ethoxy]-4,5-methylenedioxy-2-propyltoluene

SESAMEX

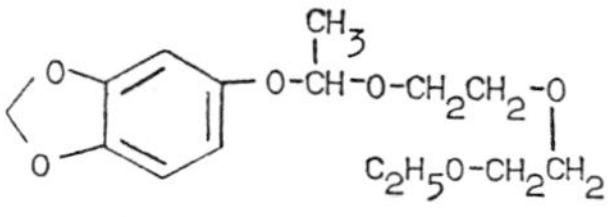

2-(2-ethoxyethoxy)ethyl-3,4-(methylenedioxy)phenyl
acetal of acetaldehyde

SULFOXIDE

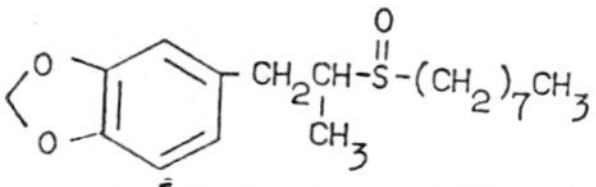

1,2-methylenedioxy-4-[2-(octylsulfinyl)propyl]benzene

MGK 264 belongs to the "no-name" moiety and appeared in 1944. It has been used in great quantity mostly in livestock and animal shelter sprays.

MGK 264®

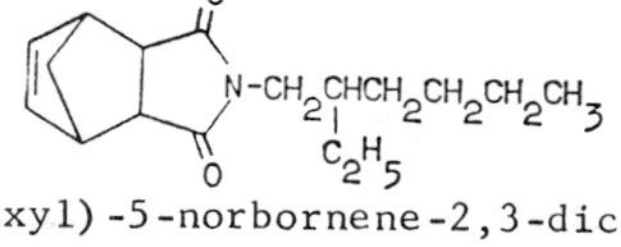

N-(2-ethylhexyl)-5-norbornene-2,3-dicarboximide

In summary, the synergists are used in many of the insecticide mixtures for home, garden, and barn. Their mode of action is the binding to oxidative enzymes that would otherwise degrade the insecticide.

5-L THE INORGANICS

The inorganic insecticides are those which do not contain carbon. Usually they are white and crystalline, resembling the salts. They are stable chemicals, do not evaporate, and are frequently soluble in water.
Sulfur is very likely the oldest known effective

insecticide. Sulfur and sulfur candles were burned by our
great-grand parents for every conceivable purpose, from
bedbug fumigation to the cleansing of a house following a
death within or removal from medical quarantine of small-
pox. Sulfur is most useful in integrated pest management
programs where target pest specificity is important. Sulfur dusts are especially toxic to mites of every variety,
such as chiggers and spider mites, thrips, newly-hatched
scale insects, and as a stomach poison for some caterpillars. Sulfur dusts and sprays are also fungicidal,
particularly against powdery mildew. At high temperatures
($>90^{\circ}$F.) sulfur can be phytotoxic.

5.64 Why is sulfur a good insecticide/miticide to use in
     integrated pest management programs?

- -

     Usually sulfur is pest-specific, leaving predators
     and parasites.

     Other inorganic materials have been used as insecti-
cides. These include compounds of mercury, boron, thallium, arsenic, antimony, selenium, and fluoride. Except for
the fluoride compounds all the others have been discontinued
because of residue persistence, high mammalian toxicity,
and superior replacement insecticides.

5.65 The heavy metal insecticides (mercury, thallium,
     arsenic, antimony, and selenium) are no longer used
     because of _____, high _____toxicity,
     and because the newer insecticides were _____.

- -

     persistence   mammalian toxicity   superior

     The arsenicals, as well as all the other metal-derived
insecticides just mentioned, are truly stomach poisons.
They exert their toxic action only after ingestion by the
insects. The same principle applies also to warm blooded
animals including man.
     Generally, these inorganic metallic salts are con-
sidered protoplasmic poisons, in that they attack or bind
several enzymes leading to their precipitation. To be insecticidal phenomenal quantities are required to kill in
this manner as compared to present-day organic insecticides.

5.66 Stomach poisons must be _____ by the insect.

- -

      eaten or ingested

5.67 The inorganic metallic salt insecticides are con-
     sidered _____ poisons, because they _____
     enzymes and lead to there _____.

- -

      protoplasmic   bind   precipitation

     The arsenicals have a rather complex mode of action.
First, they uncouple oxidative phosphorylation (by sub-
stitution of the arsenite ion for phosphorous), a major
energy-producing step of the cell. Second, the arsenate
ion inhibits certain enzymes that contain sulfhydryl (-SH)
groups. And finally, both the arsenite and arsenate ions
coagulate protein by causing the shape or configuration
of proteins to change.

5.68 Arsenates inhibit enzymes containing the _____
     group.

- -

      sulfhydryl or -SH

     The arsenicals were employed extensively from 1930 un-
til 1956, as we were making the transition from the simple
to the complex synthetic molecules. They were, in fact,
responsible for the initiation of large-scale pesticide
applications eventually leading to the intensive use of
fungicides and herbicides in modern agriculture.
     The fluorine insecticides also included organic
fluorine compounds, but these were of little importance
and seldom used. The inorganic fluorides were sodium
fluoride, used for cockroach and ant control around the
home, and barium fluosilicate, sodium silicofluoride and
cryolite (NaF; $BaSiF_6$; Na_3SiF_6; and Na_3AlF_6, respectively).
The last three were used for a time in plant protection.
     Cryolite (sodium fluoaluminate) has returned as a very
useful insecticide, under the name Kryocide[R], in integrated
pest management programs. The fluorine in cryolite acts
as a stomach insecticide and is thus effective only against
insects with chewing mouthparts. Its uses are best recognized on fruits and vegetables and the product is also
available for home gardeners.
     The mode of action of the fluoride ion is the inhibi-
tion of many enzymes that contain iron, calcium and

magnesium. Several of these enzymes are involved in energy
production in cells, as in the case of phosphatases and
phosphorylases.

5.69 Fluorides exert their action by inhibiting enzymes
      containing _____, _____, and _____.

- -

      iron, calcium and magnesium

      Boric acid (H_3BO_4), used as an insecticide against
cockroaches and other crawling household pests in the 1930s
and '40s, has returned in the 1980s. It is highly adver-
tised as "safe, does not evaporate, and continues to kill
for years". Actually, these claims are truthful. When
boric acid is very finely ground, as a dust, it is a most
remarkable insecticide, acting as both a stomach poison and
cuticle wax absorber, mentioned below in the silica gels.
      One more group of inorganics is the silica gels or
silica aerogels. These are light, white, fluffy silicates
used for household insect control. The silica aerogels
kill insects by absorbing waxes from the insect cuticle
permitting the continuous loss of water from the insect
body. The insects then gradually become desiccated and
die from dehydration. These include Dri-Die[R], Drianone[R]
and Drione[R]. The latter two are fortified with pyrethrum
and synergists, which enhance their effectiveness.

5.70 The silica aerogels remove cuticular _____
      which permit the loss of body _____.

- -

      waxes   water

5-M FUMIGANTS

      Fumigants are usually small, volatile, organic mole-
cules that are gases at temperatures above 40°F. They are
normally heavier than air, though not always, and usually
contain one or more of the halogens (Cl, Br, or F). Most
are highly penetrating, reaching through large masses of
material. They are used to kill insects, insect eggs,
and certain microorganisms in buildings, warehouses, grain
elevators, soils, greenhouses, and in packaged products
such as dried fruits, beans, grains and breakfast cereals.

5.71 Fumigants are small, _____molecules usually
     containing one or more of the _____.

----

     volatile   halogens

        Fumigants, as a group, are *narcotics*. Their mode of
action is more physical than chemical. The fumigants are
liposoluble (fat-soluble); they have common symptomology;
their effects are reversible; and their activity is altered
very little by structural changes in their molecules. Fumigants induce *narcosis*, sleep or unconsciousness, which in
effect is their mode of action on insects. The more common
fumigants are shown in Panel F. Few of these will retain
their registrations because the basic manufacturers are un-
willing to undergo the costs of providing additional data
required by EPA.

PANEL F
COMMONLY USED FUMIGANTS

Name	Chemical Formula or Structure
Methyl bromide	CH_3Br
Ethylene dibromide	$BrCH_2CH_2Br$
Ethylene dichloride	$ClCH_2CH_2Cl$
Hydrogen cyanide	HCN
Chloropicrin	Cl_3CNO_2
Sulfuryl fluoride (Vikane[R])	SO_2F_2
Vapam[R]	$CH_3NH\overset{S}{\overset{\|}{C}}\text{-S-Na}$
Telone[R]	$ClCH=CH-CH_2Cl$
Aluminum phosphide Phostoxin[R]	AlP
Chlorothene	CH_3CCl_3
Ethylene oxide	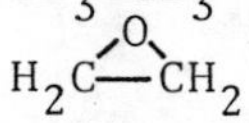 $H_2C\overset{O}{\overset{\diagdown\diagup}{-}}CH_2$
Naphthalene (crystals)	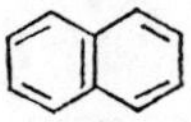
para-dichlorobenzene (PDB crystals)	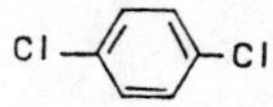

DBCP (Nemagon[R]), dibromochloropropane; D-D[R] soil fumigant, a
mixture of 1,3-dichloropropene, 1,2-dichloropropane, 3,3-di-
chloropropene, and 2,3-dichloropropene; and carbon disulfide/
carbon tetrachloride grain and soil fumigant are either no
longer registered for use or their sale and distribution
have been discontinued at the end of 1985.

5.72 The action of fumigants is _____
     and is due more to _____ than chemical
     effects.

- -

     narcotic   physical

It appears that liposolubility is an important factor in the action of fumigants, since these narcotics lodge in lipid-containing tissues, which are found in the nervous system. There are a few of the fumigants whose mode of action is more than narcotic. These are ethylene dibromide, hydrogen cyanide, and chloropicrin. Because of its strong "tear gas" effect on man, chloropicrin is commonly added in trace quantities to many fumigants as an olfactory indicator or warning gas. Otherwise many of the fumigants would be lethal to man before he became aware of a dangerous concentration in the air breathed.

5.73 Many fumigants have a _____
     added to indicate to the user a dangerous air concentration of the fumigant.

- -

     warning gas

5-N MICROBIALS

The microbial insecticides obtain their name from microbes, or *microorganisms*, which are used by man to control certain insects. Insects, like mammals, also have diseases caused by fungi, bacteria, and viruses. In several instances these have been isolated, cultured, and mass-produced for use as insecticides.

5.74 The microbials are _____ _____
     which have been isolated and used as insecticides.

- -

     disease microorganisms

Microorganisms causing disease in insects do not harm other animals or plants. The reverse of this is also true. The method of insect control is ideal in that the diseases are usually rather specific. Undoubtedly the future holds many such materials in the arsenal of insecticides, since several new insect pathogens are identified each year. However, at the present only a few are produced commercially and approved by EPA for use on food and seed crops. There is still some concern regarding the very remote chance of human susceptibility to these insect diseases, thus the slow advances into this relatively new field and exceptional precautionary testing.

Bacillus thuringiensis is a disease-causing bacterium, whose spores are necessary for disease induction. These

spores produce compounds that injure the gut of insect
larvae in a way that invasion of the body cavity follows.
This organism produces four substances toxic to insects.
The first, and most important, is a crystalline protein
that after ingestion results in a paralysis of the gut of
caterpillars. The second is a toxin, a water soluble
nucleotide derivative, that will pass unchanged through
the gut of a cow and kill certain fly maggots breeding in
the manure. The remaining two substances are enzymes
which are lost in the commercial preparation. The structures of the active agents are not known.
      Bacillus thuringiensis var. kurstaki microbial insecti-
cides are marketed under several trade names, of which
Dipel^R, Sok-Bt^R, Thuricide^R, Certan^R, and Bactospeine^R are
examples. The most recent addition to the "b-ts" is
Javelin^R introduced in 1985, with increased activity against
the beet armyworm.

5.75 *Bacillus thuringiensis* spores produce products which
     act mainly in the _____ of insect larvae.

- -

     gut

      Bacillus thuringiensis var. israelensis is marketed
under the trade names Bactimos^R, BMC^R, Teknar^R and
Vectobac^R. This strain is most effective against mosquito
and black fly larvae, the active ingredient being delta-
endotoxin in crystalline form. Two slow-release formulations are available for mosquito control (Bactimos Briquets^R
and Mosquito Attack Rings^R).
      One insect virus known as the *Heliothis* nuclear poly-
hedrosis virus has become registered for agricultural use.
It is specific for *Heliothis zea* (corn earworm, cotton bollworm) and *Heliothis virescens* (tobacco budworm), two of the
most destructive pest caterpillars in agriculture. Recently
registration has been extended not only for cotton, but for
all crops attacked by *Heliothis* species, including beans,
corn, lettuce, okra, peppers, sorghum, soybeans, tobacco,
and tomatoes. The proprietary name for this naturally occurring viral pathogen is Elcar^R.
      Other insect viruses are in the development stage, such
as the granulosis viruses, one of which is in experimental
development for control of the codling moth. Viruses are
highly specific and have modes of action that may not be
identical throughout. Generally, the viruses result in crystalline proteins that are eaten by the larva and begin
their activity in the gut. The virus unit then passes
through the gut wall and into the blood. From there, several possibilities occur, but the known aspects are that the
units multiply rapidly and take over complete genetic control of the cells, causing their death.

5.76 Viruses multiply rapidly in the insect and take
_____ control of the cells resulting in death.

- -
     genetic

_____

     An advance in the agricultural use of microbials is
the addition of feeding or gustatory stimulants which
attract the caterpillars to treated foliage, resulting in a
heavier dose of the microbial than without the stimulant.
Two successful products are Coax^R and Gustol^R. Both are
water soluble and are used at 1 to 2 lb per acre.
     More recently, Bollex^R, a boll weevil feeding deter-
rent, has become available. It is specific for this cotton
pest, discouraging weevils from feeding. Bollex^R is not
truly an insecticide in that it does not kill weevils.
     A parasitic fungus that kills the citrus rust mite is
registered under the name Mycar^R. It is the first myco-
acaricide, developed from the fungus *Hirsutella thompsonii*.
Under ideal conditions it also infects spider mites and
other non-target mites. However, it is consistently effective and selective against the citrus rust mite.
     Another microbial insecticide is the protozoan, *Nosema
locustae,* developed for control of grasshoppers and crickets.
It is marketed under the names NOLOC, Cricket Attack,
Grasshopper Attack, Hopper Stopper, and Mormon Cricket
Spore, among others. This biological insecticide is most
effective applied as a bait and its greatest usefulness is
in the control of rangeland grasshoppers or locusts.
     The first nematode to be registered for insect control
is *Neoplectana carpocapsae,* marketed under the names of
Spear^R and Saf T-shield^R. It is specific for termites and
kills all stages by delivering a pathogenic bacterium lethal
to these wood-eating pests 48 hours after penetration.

_____

5.77 Biological insecticides are ideally suited for in-
     clusion in integrated pest management programs. Why?

_____

- -

     They are pest specific.

_____

5-O   ANTIBIOTICS

     Antibiotics, as you probably know them, are those such
as penicillin, tetracycline, and chloramphenicol, used
against bacterial diseases of man and his domestic animals.
These are not involved in our present study. However, a
new insecticide-miticide, Avermectin^R, is derived from the
antibiotic-producing actinomycetes, in this case *Streptomyces avermitilus.* The actinomycetes are the source of all
the antibiotic fungicides, as you will read later.

Avermectin^R is actually a mixture of two homologs, avermectin Bla and Blb, which have equal biological activity. Avermectin is active against the two major groups, the nematodes and arthropods (insects, ticks and mites). Its greatest promise is in the control of mites on citrus, especially the citrus rust mite and the two spotted mite. Certain insects are also controlled, including leafminers on greenhouse chrysanthemums, field celery and tomatoes and the cotton leafperferator.

Avermectin^R

Its mode of action is the blocking of the neurotransmitter gamma aminobutyric acid (GABA), at the neuromuscular junction in insects and mites. Visible activity, such as feeding and egg laying, stops shortly after exposure, though death may not occur for several days.

Avermectin^R has certain local systemic qualities, permitting it to kill mites on a leaf's underside when only the upper surface was treated. This quality also explains its success in controlling leaf miners and the cotton leafperferator.

5.78 A new mode of action described for the first antibiotic
     insecticide is the _____ of the neurotransmitter
     GABA at the _____ junction.

- -

     blocking   neuromuscular

5-P   INSECT GROWTH REGULATORS

Insect growth regulators (IGRs) are the ultimate in insecticides. The first generation insecticides were the stomach poisons, such as the arsenicals. The second generation includes the familiar contact insecticides: organochlorines, organophosphates, carbamates and formamidines. We are now on the edge of the third generation, the *biorationals*. These are chemicals which resemble closely or are identical to chemicals produced by insects and plants, which are environmentally sound, and can be used rationally in their management and control. Among these

biorationals are the IGRs.
  IGRs are a new group of compounds that alter growth
and development in insects. Their effects have been
observed on all stages of growth and metamorphosis, on
reproduction, behavior, and diapause. They include the
molting and juvenile hormones and more recently chitin
inhibition.
  The IGRs are required only in ultra-low quantities
and have no undesirable effects on humans and wildlife.
Because they are nonspecific, however, they affect not
only the target species but most other arthropods as well.
  There are four IGRs registered by the EPA. The first,
methoprene, was accepted in 1975 as Altosid^R for floodwater
mosquito control, used at 0.1 to 0.12 lb per acre to prevent
adult emergence. Other formulations include Apex^R for
sciarid fly control in mushrooms, Diacon^R for stored peanut
pests, Dianex^R for stored products pests in tobacco and
food plants, Kabat^R for stored tobacco pests, Minex^R for
leafminers in chrysanthemum greenhouses, Pharorid Bait^R for
pharoah ant control, and Precor^R for indoor flea control.
  Next is diflubenzuron (Dimilin^R), which is not truly a
juvenile hormone regulator. Rather it acts on the larval
stages of insects by blocking the synthesis of chitin, the
hard outer covering. Currently it is registered for gypsy
moth and several other forest caterpillars, the cotton boll
weevil, velvetbean caterpillar and green cloverworm on soybeans, and fly larvae in mushroom growing facilities.
  The third is kinoprene (Enstar^R). It is specific for
insects in the order Homoptera, and is effective against
aphids, whitefly, mealybug and scales on selected greenhouse
and shadehouse crops.
  The fourth and most recent is hydroprene, marketed as
Gencor^R (formerly Altozar^R). It is used in the control of
cockroaches by causing them to die before becoming reproductive adults. Though not intended for their control,
hydroprene is also active against homopteran, lepidopteran
and coleopteran species. (Not illustrated)
  In summary, IGRs hold intriguing possibilities for
future use in practical insect control. It should be kept
in mind that IGRs are insect-controlling chemicals and thus
fall within the same legal confines as other insecticides.
Their great distinction, however is that they are very
selective and are harmless to warm blooded animals.

5.79 What are the 3 generations of insecticides? Give an
      example of each. 1. _____
      2. _____
      3. _____

      Stomach poisons (arsenicals)
      Contact insecticides (organochlorines, organophosphates)
      Biorationals (IGRs)

METHOPRENE (Altosid[R])

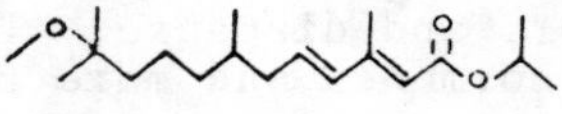

isopropyl (2E-4E)-11-methoxy-3,7-
11-trimethyl-2,4-dodecadienoate

DIFLUBENZURON (Dimilin[R])

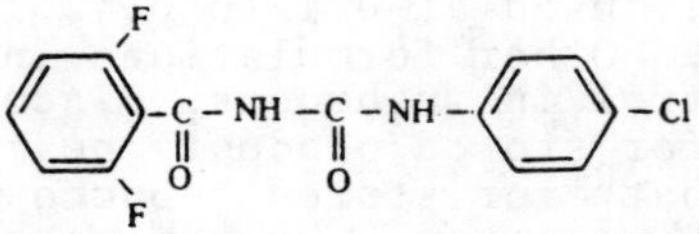

1-(4-chlorophenyl) 3-(2,6-difluorobenzoyl)urea

KINOPRENE (Enstar[R])

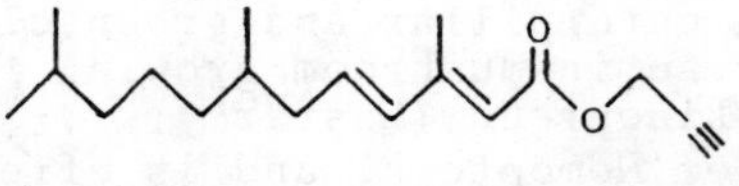

2-propynyl (E,E)-3,7,11-trimethyl-2,4-dodecadienoate

5-Q REPELLENTS

      Prior to World War II, there were only four reasonably
effective repellents: Oil of citronella, dimethyl phthalate, indalone and Rutgers 612. With the introduction of
American military personnel into tropical environments
it became necessary to find new repellents that would be
long lasting and survive dilution by perspiration. Ideally
they would be nontoxic and nonirritating to humans, non-
plasticizing and last at least 12 hours against chiggers,
fleas, ticks, biting flies and mosquitoes. The ideal
repellent has still not been found.
      Following are shown the chemical structures of the
most commonly used repellents. Of these, diethyl toluamide
(Delphene[R], deet), introduced in 1955, is still superior to
all others against biting flies and mosquitoes.

DEET (Delphene^R)

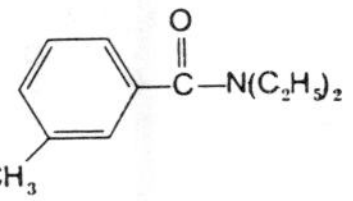

N,N-diethyl-m-toluamide

INDALONE

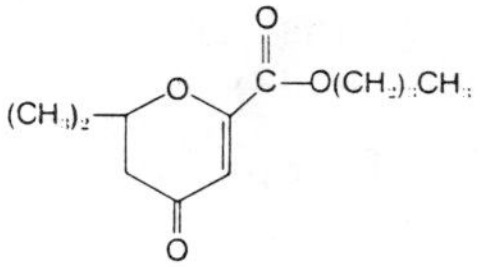

butyl 3,4-dihydro-2,2-dimethyl-
-4-oxo-2H-pyran-6-carboxylate

MGK REPELLENT 326

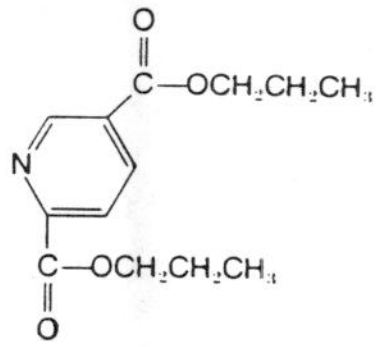

di-n-propyl 2,5-pyridinedicarboxylate

DIBUTYL PHTHALATE

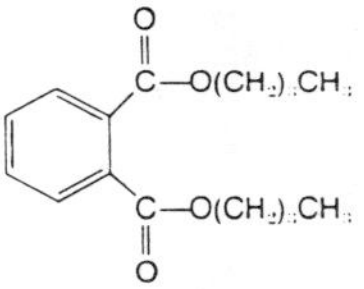

di-n-butyl phthalate

RUTGERS 612

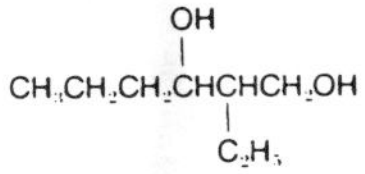

2-ethyl-1,3-hexanediol

CLOTHING IMPREGNANT (ticks, chiggers)

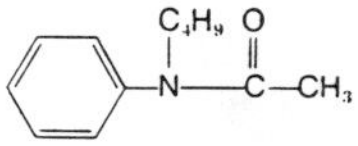

N-butyl acetanilide

MGK REPELLENT 11

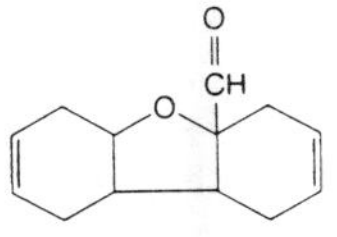

1,5a,6,9a9β-hexahydro-4a(4H)-
dibenzofuran carboxaldenyde

DIMETHYL CARBATE (Dimelone^R)

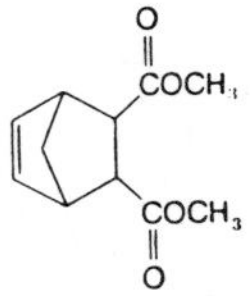

dimethyl cis-bicyclo(2,2,1)-5-
heptane-2,3-dicarboxylate

5.80 Of the many insect repellents that have appeared over
     the years, only one has really been outstanding and
     genuinely effective. That is _____.

- -

     deet or diethyl toluamide

5.81 How is your memory? As a quick review, see if you
can identify the insecticide structures given below:
organosulfur_____; organochlorine_____; organo-
phosphate_____; carbamate_____; synthetic pyrethroid
_____; synergist_____; botanical_____; fumigant_____;
dinitrophenol_____.

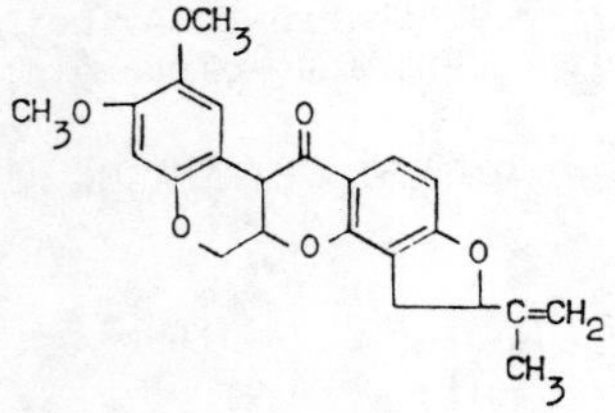

1

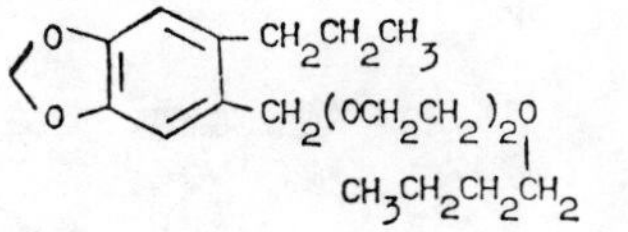

2

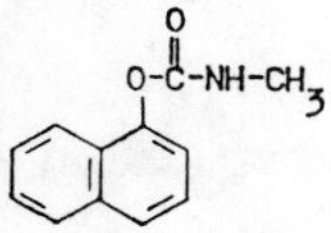

BrCH₂CH₂Br

4

3

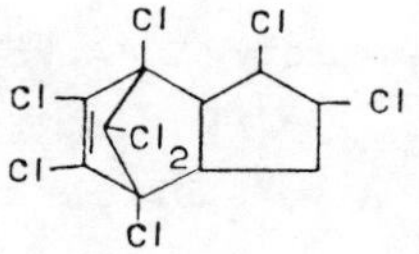

5

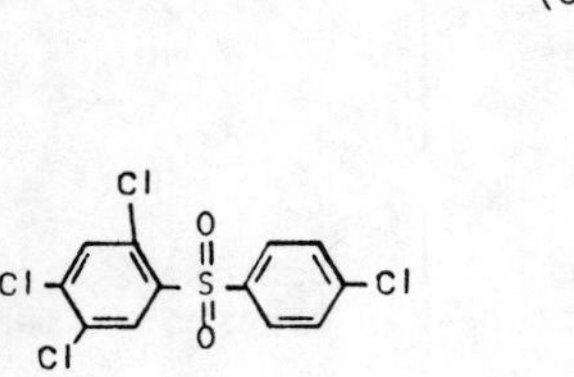

6

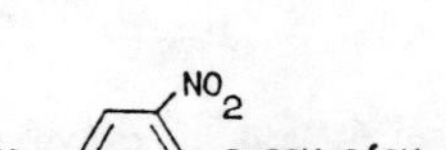

7

8

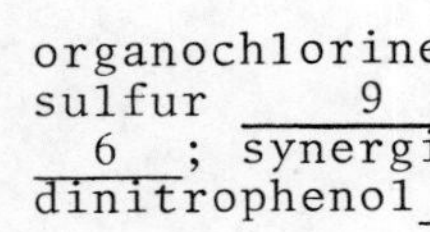

9

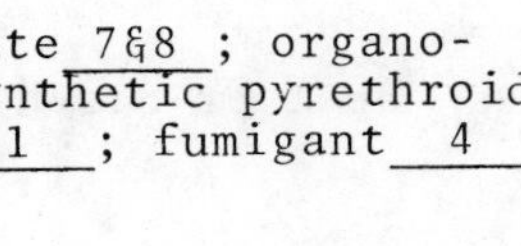

10

- -

organochlorine   5   ; organophosphate 7&8 ; organo-
sulfur   9   ; carbamate   3   ; synthetic pyrethroid
   6   ; synergist   2   ; botanical   1   ; fumigant   4   ;
dinitrophenol   10   .

UNIT 6
HERBICIDES

In the past 40 years, chemical weedkillers, or herbicides, have largely replaced mechanical methods of weed control, especially where intensive and highly mechanized farming is practiced. Herbicides provide a more effective means of weed control than cultivation, hoeing, and hand pulling. Together with fertilizers, other pesticides, and improved plant varieties, they have made an important contribution to the increased yields we now have, and are one of the few ways remaining to combat rising costs and shortage of agricultural labor. Heavy use of herbicides is confined to North America, Western Europe, Japan, and Australia. Without the use of herbicides, it would have been impossible to mechanize fully the production of grains, sugar beets, potatoes, corn, and cotton.

Herbicides are also used extensively in other locations. These include areas such as industrial sites, roadsides, ditch banks, irrigation canals, fence lines, recreational areas, railroad embankments, and power lines. Herbicides remove undesirable plants that might cause damage, present fire hazards, or impede work crews. They also reduce costs of labor for mowing.

Herbicides are classed as *selective* when they are used to kill weeds without harming the crop, *non-selective* when the purpose is to kill all vegetation.

6.1 Herbicides have many non-agricultural uses. Cite several examples: _____, _____, _____, _____, and _____.

- -

industrial sites, roadsides, ditch banks, canals, fence lines, recreational areas, etc.

6.2 Herbicides are broadly classed into the _____ and _____ categories.

- -

selective and non-selective

Depending on the mode of action, both selective and
non-selective materials can be applied to weed foliage
or to soil containing weed seeds and seedlings.
    True selectivity refers to the capacity of an herbi-
cide, when applied at the proper dosage and time, to be
active only against certain species of plants but not
against others. But selectivity can also be achieved by
placement, as when a non-selective herbicide is applied
in such a way that it contacts the weeds but not the crop.

6.3 Herbicide selectivity can be either _____
     _____ or it can be achieved by _____.

- -

     true selectivity   placement

    The classification of herbicides would be a simple
matter if only the selective and nonselective categories
existed. But it is difficult to explain the multiple-
classification schemes, which may be based on *selectivity,
contact vs. translocated, timing, area covered,* and
chemical classification.
    All herbicides affect plants by contact or translo-
cation. The *contact herbicides* kill the plant parts to
which the chemical is applied. They are most effective
against *annual weeds,* those that germinate from seeds and
grow to maturity each year. Complete coverage is essential in weed control with contact materials.

6.4 Contact herbicides kill only that part of the plant
     which is _____, are most effective
     against _____ weeds, and require _____
     coverage of weeds.

- -

     treated   annual   complete

    Translocated herbicides are absorbed either by roots
or above-ground parts of plants and then moved within the
plant system to distant tissues. *Translocated herbicides*
may be effective against all weed types, and their greatest advantage is seen when used to control established
perennial weeds, those that continue their growth from year
to year. Uniform application is needed for the translocated materials, whereas complete coverage is not
required.

6.5 Why would translocated herbicides be used to greatest
     advantage in controlling perennial weeds?

_____

_____

- -

     Through systemic movement the herbicide is carried
     to the roots, which survive from year to year in
     perennials.

6.6 Translocated herbicides are _____
     from the point of application to other plant parts,
     are most useful against _____weeds,
     require _____application, but not
     _____ coverage.

- -

     moved   perennial   uniform   complete

_____

     The timing of herbicide application with regard to
the stage of crop or weed development is another method of
classification. Applications may depend on many factors,
including the chemical classification of the material and
its persistence, the crop and its tolerance to the herbicide, weed species, cultural practices, climate, and soil
type and condition. The three categories of timing are
preplanting, preemergence, and *postemergence.*
     Preplanting applications are applied to an area
before the crop is planted, within a few days or weeks of
planting for control of annual weeds. Preemergence applications are completed prior to emergence of the crop or
weeds, depending on definition, after planting. Postemergence applications are made after the crop or weed
emerges from the soil.

_____

6.7 Timing of herbicide applications may be _____,
     _____ or _____.

- -

     preplanting, preemergence, postemergence

_____

     The classification based on area covered during the
herbicide application involves four categories: *band,
broadcast, spot treatments,* and *directed spraying.* A band
application treats a continuous strip, as along or in a
crop row. Broadcast applications cover the entire area,
including the crop . Spot treatments are confined to small
areas of weeds. Directed sprays are applied to selected
weeds or to the soil to avoid contact with the crop.

_____

6.8 Based on the area covered, there are four methods of
     applying herbicides: _____, _____,
                          _____, and _____ _____.
- -

     band   broadcast   spot   directed spraying

     This brings us to the chemical classification of her-
bicides, where the major emphasis of this book is placed.
The two major classifications are inorganic and organic
herbicides.

INORGANIC HERBICIDES

     Inorganic compounds were the first chemicals utilized
in weed control. These were brine, and a mixture of salt
and ashes, both of which were used as early as Biblical
times. In 1896, copper sulfate was used selectively to
kill weeds in grain fields. From about 1906 until 1960,
sodium arsenite solutions were the standard herbicides of
commerce. Arsenic trioxide has been used at the incredi-
ble rates of 400 to 800 pounds per acre for soil sterili-
zation. As with the arsenical insecticides, the trivalent
arsenicals were nonspecific inhibitors of those enzymes
containing sulfhydryl groups. They also uncouple oxidative phosphorylation.

6.9 The arsenical herbicides act by _____
                          _____-containing enzymes and
     uncoupling _____ _____.
- -

     inhibiting   sulfhydryl or -SH
     oxidative phosphorylation

     Ammonium sulfamate ($NH_4SO_3NH_2$) was introduced in
1942 for brush control. Other salts have been used over
the years. These include ammonium thiocyanate, ammonium
nitrate, ammonium sulfate, iron sulfate, and copper sulfate, each applied heavily as a foliar spray. Their mechanisms of action are desiccation and plasmolysis (shrink-
age of cell protoplasm away from its wall due to removal
of water from its large central vacuole).

6.10 These miscellaneous salts act against plants by
     _____ and _____.

- -

     plasmolysis   desiccation

Another family of inorganics was the borate herbicides, for example, sodium tetraborate ($Na_2B_4O_7 \cdot 5H_2O$), sodium metaborate ($Na_2B_2O_4 \cdot 4H_2O$), and amorphous sodium borate ($Na_2B_8O_{13} \cdot 4H_2O$). The amount of boron or boric acid determined their effectiveness. Borates are absorbed by plant roots, translocated to above-ground parts, and are nonselective, persistent herbicides. Boron accumulates in the reproductive structures of plants, but its mechanism of toxicity is unclear.

Sodium chlorate ($NaClO_3$) has been used as a nonselective herbicide for the last 40 years. It acts as a soil sterilant at rates of 200 to 1000 pounds per acre, but it can be used as a foliar spray at 5 pounds per acre as a defoliant of cotton. Caution must be taken with sprays of sodium chlorate to be certain the formulation contains flame retardants. Sulfuric acid has also been used as a foliar herbicide, but its use is limited considerably by its corrosiveness to metal spray rigs. Their mechanisms of action are those described above for the miscellaneous salts — desiccation and plasmolysis.

The inorganic herbicides are still useful in weed and brush control, but are gradually being replaced by organic materials. Though organic herbicides are not superior to inorganic ones, intensive EPA restrictions have been placed on inorganic herbicides because of their persistence in soils. Inorganic herbicides should not be used around the home except by professionals, and then only when it is desired that all vegetation be removed. Weeds and eventually trees respond alike.

ORGANIC HERBICIDES

6-A PETROLEUM OILS

The first of the organic herbicides were the petroleum oils, which are a complex mixture of long-chain hydrocarbons containing traces of nitrogen- and sulfur-linked compounds. These mixtures include alkanes, alkenes, and often alicyclics and aromatics, from distillation and refining of crude oils. The petroleum oils are effective contact herbicides for all vegetation. Homeowners today use old crank case oil, gasoline, kerosene and diesel oil for spot treatments. The petroleum oils exert their lethal effect by penetrating and disrupting plasma membranes. Other oils used in agriculture are naptha, Stoddard solvent, fuel oils, and mineral spirits. These materials are fast-acting and safe to use around the home, but yield only temporary results.

6.11 Petroleum oils act by disrupting the _____
     _____.

- -

     plasma membranes

6-B ORGANIC ARSENICALS

The arsinic and arsonic acid derivatives are widely used as herbicides. Cacodylic acid (dimethylarsinic acid) and its sodium salt are the only derivatives of arsinic acid. Salts of arsonic acid are disodium methanearsonate (DSMA), and monosodium acid methanearsonate (MSMA).

ARSONIC ACID	ARSINIC ACID

$$\text{ARSONIC ACID}\qquad R-\overset{\displaystyle O}{\underset{\displaystyle OH}{\overset{\|}{As}}}-OH$$

$$\text{ARSINIC ACID}\qquad R-\overset{\displaystyle O}{\underset{\displaystyle R'}{\overset{\|}{As}}}-OH$$

MSMA

$$CH_3-\overset{\displaystyle O}{\underset{\displaystyle OH}{\overset{\|}{As}}}-ONa$$

monosodium methanearsonate

CACODYLIC ACID

$$CH_3-\overset{\displaystyle O}{\underset{\displaystyle CH_3}{\overset{\|}{As}}}-OH$$

hydroxydimethylarsine oxide

DSMA

$$CH_3-\overset{\displaystyle O}{\underset{\displaystyle ONa}{\overset{\|}{As}}}-ONa$$

disodium methanearsonate

CACODYLIC ACID

$$CH_3-\overset{\displaystyle O}{\underset{\displaystyle CH_3}{\overset{\|}{As}}}-ONa$$

sodium salt

All of the organic arsenicals are crystalline solids, relatively soluble in water, and much less toxic to mammals than the inorganic forms. The arsonate, or pentavalent arsenic, acts in a way different from the trivalent form just described in the inorganic arsenicals. The arsonates upset plant metabolism and interfere with normal growth by entering into reactions in place of phosphate. Not only do they substitute for essential phosphate, but the arsonates are absorbed and translocated in a manner similar to the absorption and translocation of phosphates.

6.12 The arsonates act by substituting for _____
     in normal plant metabolism.

- -

     phosphate

6.13 The arsenites act by _____
     and _____ (See page 91) .

     ----

     inhibiting -SH-containing enzymes and
     uncoupling oxidative phosphorylation

6.14 The organic arsenical herbicides are derivatives of
     _____ and _____ acids.

     ----

     arsinic   arsonic

6-C PHENOXYALIPHATIC ACIDS

     In 1944, an organic herbicide was introduced, later
known as 2,4-D, the first of the "phenoxy herbicides,"
"phenoxyacetic acid derivatives," or "hormone" weed killers.
These were highly selective for broadleaved weeds and were
translocated throughout the plant. 2,4-D provided most of
the impetus in the commercial search for other organic
herbicides in the 1940s. There are several compounds in
this group, of which 2,4-D and 2,4,5-T are the most familiar.
Other compounds in this group are 2,4-DB, MCPA, and silvex.
All registrations for 2,4,5-T and silvex have been cancelled.

2,4-D

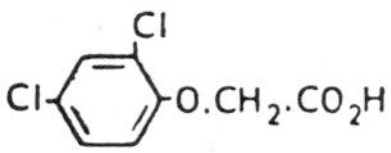

Cl

Cl—⟨ ⟩—O.CH$_2$.CO$_2$H

(2,4-dichlorophenoxy)acetic acid

2,4,5-T

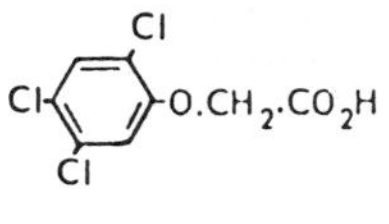

Cl

Cl—⟨ ⟩—O.CH$_2$.CO$_2$H
       |
       Cl

(2,4,5-trichlorophenoxy)acetic acid

2,4-DB

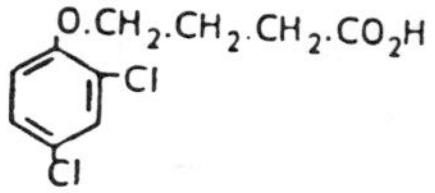

O.CH$_2$.CH$_2$.CH$_2$.CO$_2$H

Cl
|
Cl

γ-(2,4-dichlorophenoxy) butyric acid

MCPA

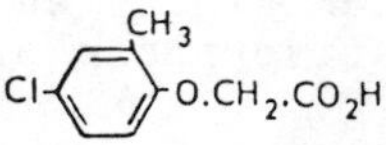

$$Cl-\underset{CH_3}{\bigcirc}-O.CH_2.CO_2H$$

[(4-chloro-<u>o</u>-tolyl)oxy]acetic acid

6.15 The phenoxy herbicides are best represented by those
twins, _____ and _____ .

- -

2,4-D and 2,4,5-T

From their introduction, 2,4-D, 2,4,5-T and MCPA were
used in gargantuan volume world-wide with no known adverse
effects on human or animal health. However, 2,4,5-T, used
for the control of woody perennials, came under heavy investigation, particularly with its use in Vietnam, as a component of Agent Orange. Excessive amounts of a highly toxic
impurity, tetrachlorodioxin, commonly referred to as dioxin,
were found in certain samples. Alterations in manufacturing
processes brought the dioxin content to tolerable levels
after the source was determined. All uses for 2,4,5-T have
since been cancelled by EPA following Dow Chemical Company's
voluntary cancellation of its registration.

The old standby, 2,4-D, continues to be one of the most
useful herbicides developed. It is used in the form of
1,500 products, in 35 ester and salt forms, amounting to
more than 70 million pounds annually, manufactured by six
basic producers in the U. S. alone. Its manufacturing process in the U. S. does not result in dioxin contamination,
though processes used in other countries sometimes do.

The modes of action of the phenoxyaliphatic acid herbicides are complex, resembling those of *auxins* (growth hormones) in some. These herbicides affect cellular division,
activate phosphate metabolism, and modify nucleic acid
metabolism.

6.16 What was the impurity in the manufacture of 2,4,5-T?

- -

tetrachlorodioxin or dioxin

6.17 The actions of phenoxy herbicides resemble those of
     plant _____ and affect cellular _____.
----

     hormones (auxins)   division

6.18 The phenoxyaliphatics were the first highly
     _____ and _____ herbicides.
----

     selective   translocated

6-D SUBSTITUTED AMIDES

     The biological properties of the amide herbicides
are diverse. These are simple molecules that are easily
degraded by plants and soil. One of the earliest was CDAA
(allidochlor), which is selective for the grasses. Allidochlor inhibits the germination of early seedling growth
of most annual grasses probably by alkylation of the -SH
groups of proteins. Production of CDAA was discontinued in
1984.

CDAA (Randox[R])

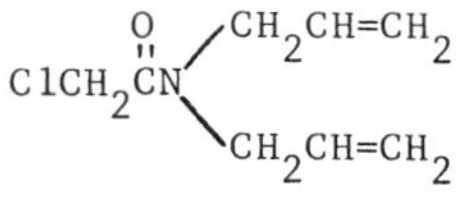

N,N-diallyl-2-chloroacetamide

     Diphenamid is used as a preemergence soil treatment,
and has little contact effect. Most established plants
are tolerant to diphenamid, because it affects only seedlings. It persists from 3 to 12 months in soil.

DIPHENAMID (Dymid[R], Enide[R])

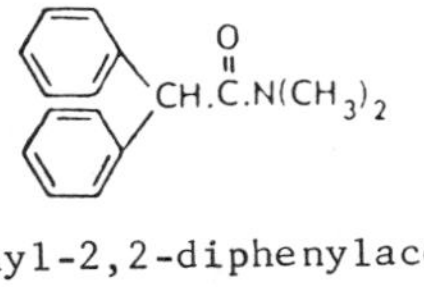

N,N-dimethyl-2,2-diphenylacetamide

Propanil has been used extensively on rice fields as a selective postemergence control for a broad spectrum of weeds.

PROPANIL

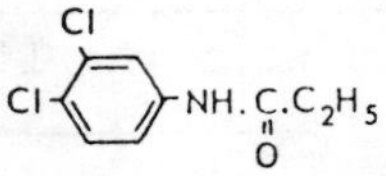

3',4'-dichloropropionanilide

Napropamide controls grass and broadleaf weeds in vineyards and orchards, and in direct seeded tomatoes, strawberries, tobacco, and ornamentals.

NAPROPAMIDE (Devrinol[R])

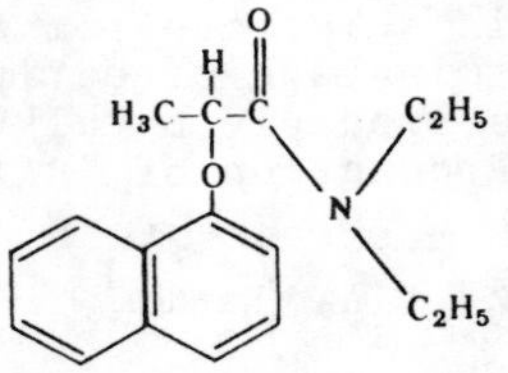

2-(a-naphthoxy)N,N-
diethylpropionamide

Propanil acts primarily in the leaves and is a strong inhibitor of the *Hill reaction*. This is a light-initiated reaction that splits water (photolysis), resulting in the production of free oxygen (O_2) by plants. Chlorophyll, the green pigment of plants, is an essential ingredient in the reaction, since it catalyzes the production of oxygen from water and the transfer of the hydrogen to a hydrogen-acceptor. A simplified chemical formula for the reaction could be written as follows:

$$2H_2O + 2A \xrightarrow[\text{chlorophyll}]{\text{light}} 2AH_2 + O_2$$

where "A" is some unidentified hydrogen-acceptor. The hydrogen and acceptor complex (AH_2) continue on in reactions with CO_2 to form plant sugars and cellulose while free O_2 is released into the atmosphere.

6.19 Propanil acts by inhibiting _____
     in affected plants.

----

     the Hill reaction

6.20 The Hill reaction explains how _____ is split
     in the presence of chlorophyll and light to produce
     _____ and _____.

----

     water   oxygen and hydrogen

     Naptalam and alachlor are used as preemergence sprays
for controlling germinating seedlings of both grasses and
broad-leaved plants. Naptalam acts by blocking indoleacetic
acid (IAA, see Plant Growth Regulators), while alachlor
inhibits protein synthesis.

NAPTALAM (Alanap[R])

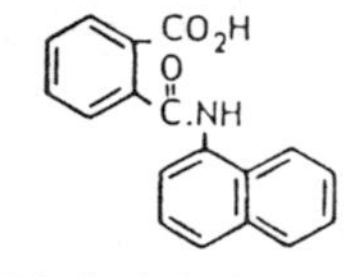

N-1-naphthylphthalamic acid

ALACHLOR (Lasso[R])

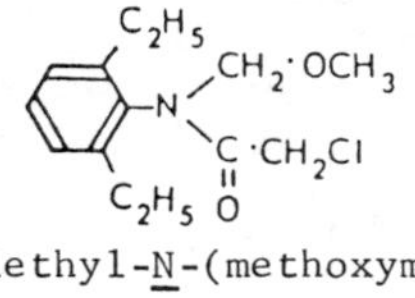

2-chloro-2',6'-diethyl-N-(methoxymethyl)acetanilide

PRONAMIDE (Kerb[R])

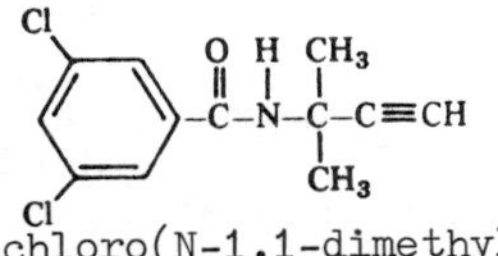

3,5-dichloro(N-1,1-dimethyl-
2-propynyl) benzamide

Metolachlor (Dual^R) was introduced in 1974, and belongs to the chloroacetamide herbicides (or acetanilides). It is soil-incorporated as a preplant and preemergence herbicide for annual weed grasses and some broad leaf weeds in corn, soybeans and peanuts. Its mode of action is probably the inhibition of nucleic acid metabolism and protein synthesis.

METOLACHLOR (Dual^R)

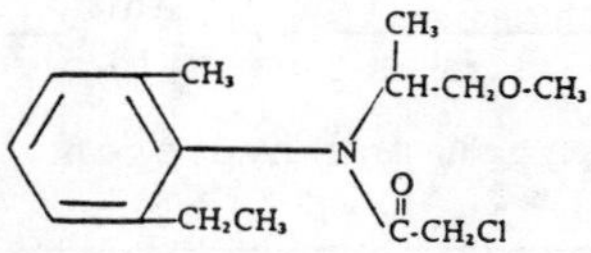

2-chloro-N-(2-ethyl-6-methylphenyl)-
N-(2-methoxy-1-methylethyl) acetamide

The substituted amide group is one of the largest, containing many other compounds not mentioned. The ones discussed and illustrated, however, are the most important of that group and will serve to provide the overview.

6-E DIPHENYL ETHERS

The diphenyl ether herbicides are constructed around a molecule made of two phenyl (benzene) rings tied together with an oxygen.

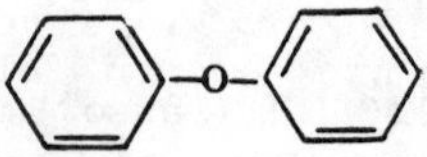

The first of this group was nitrofen, introduced in 1963. They are generally used to control annual weeds with a preemergence or early postemergence application. They are primarily contact herbicides and are readily absorbed by leaves and roots. Translocation is limited. They induce chlorosis and necrosis as visible effects of activity. Physiologically, their mode of action is not clearly understood, but thought to inhibit electron transport and coupled photophosphorylation. In some instances they induce growth responses, resembling the action of auxins and the phenoxy herbicides.

Aciflourfen (sodium salt) is a recent addition to the diphenyl ether herbicides, as is diclofop methyl. Acifluorfen controls a wide spectrum of annual broadleaf and grass weeds in soybeans, peanuts and other legumes. Diclofop methyl, sold under the name of Illoxan^R in other countries, is used for control of annual grassy weeds in wheat, barley and soybeans.

ACIFLUORFEN (Blazer[R])

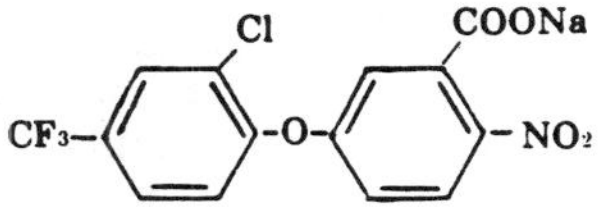

sodium 5-(2-chloro-4-(trifluoromethyl)-phenoxy)-2-nitrobenzoate

DICLOFOP METHYL (Hoegrass[R], Hoelon[R], Illoxan[R])

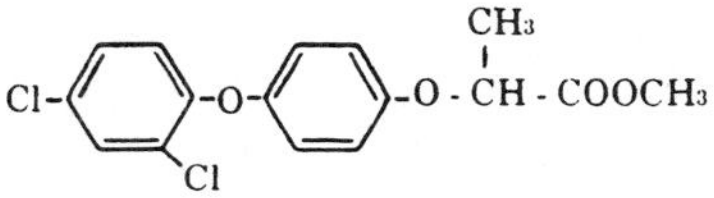

2-(4-(2,4-dichlorophenoxy-
phenoxy)-methyl-propanoate

Oxyfluorfen controls annual broadleaf and grass weeds in soybeans, corn (witchweed), cotton, deciduous fruits, grapes and nuts.

OXYFLUORFEN (Goal[R])

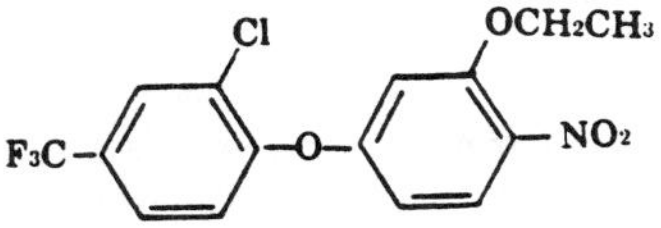

2-chloro-1-(3-ethoxy-4-nitrophenoxy)-
4-(trifluoromethyl) benzene

The most recent member is lactofen, promising as a selective postemergence compound with some preemergence activity against broadleaved weeds, is under experimental use permit for soybeans, cotton, peanuts and rice.

LACTOFEN (Cobra^R)

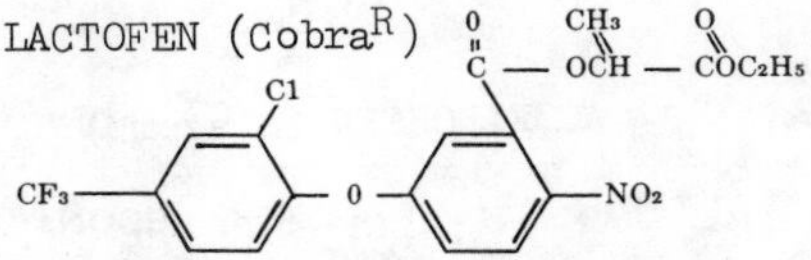

1'-(carboethoxy) ethyl 5-(2-chloro-4-
(trifluoromethyl) phenoxy)-2-nitrobenzolate

Fluazifop-butyl is a new, selective systemic herbicide, which resembles the diphenyl ether group as well as the phenoxys, but actually belongs to neither. In contrast to the diphenyl ethers, it is used exclusively for control of grass weeds in broadleaved crops. In 1985 its registrations covered only cotton and soybeans.

FLUAZIFOP-BUTYL (Fusilade^R)

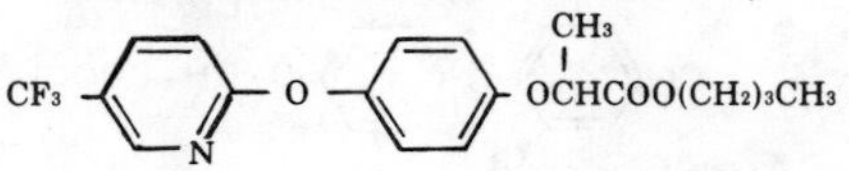

butyl 2-(4-(5-trifluoromethyl-2-
pyridyloxy)phenoxy) propionate

Fenoxaprop-ethyl, also without a home, is a postemergence, selective material used to control grasses, including Johnsongrass, and still experimental at this writing.

FENOXAPROP-ETHYL (Furore^R, Whip^R, Acclaim^R)

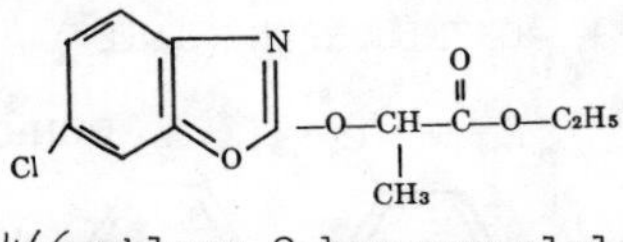

ethyl 2(4(6-chloro-2-benzoxazolyloxy)-
phenoxy) propanoate

6.21 The diphenyl ethers have a molecular structure easily recalled. Diphenyl (2 phenyl rings) with an ether (oxygen) linkage. See if you can reconstruct this structure below.

- -

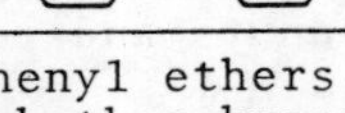

_____

6.22 The diphenyl ethers are used generally to control annual weeds, whether broadleaf or grasses. This is because they are contact herbicides and not _____.

- -

systemic

_____

6-F NITROANILINES

The nitroanilines are possibly the heaviest used group of herbicides in agriculture. They are used almost exclusively as soil-incorporated, preemergence selective herbicides in many field crops. Examples of the substituted dinitrotoluidines are trifluralin, benefin, nitralin, and isopropalin. Trifluralin has very low water solubility, which minimizes leaching and movement away from the target. The nitroanilines inhibit both root and shoot growth when absorbed by roots, but they have an involved mode of action, which includes inhibiting the development of several enzymes and the uncoupling of oxidative phosphorylation.

6.23 Nitroanalines act by both uncoupling_____
_____ and inhibiting production of
_____.

- -

oxidative phosphorylation   enzymes or proteins

TRIFLURALIN (Treflan^R)

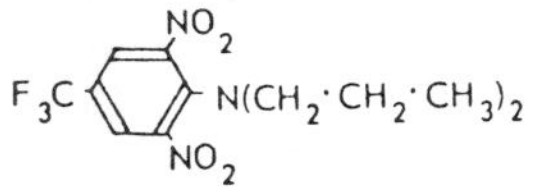

$$F_3C \underset{NO_2}{\overset{NO_2}{\bigcirc}} N(CH_2 \cdot CH_2 \cdot CH_3)_2$$

α,α,α-trifluoro-2,6-dinitro-N,N-dipropyl-p-toluidine

BENEFIN (Balan^R)

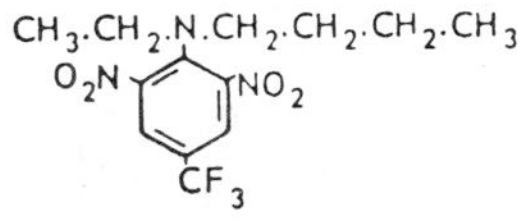

$$CH_3 \cdot CH_2 \cdot N \cdot CH_2 \cdot CH_2 \cdot CH_2 \cdot CH_3$$
$$O_2N \overset{}{\bigcirc} NO_2$$
$$CF_3$$

N-butyl-N-ethyl-α,α,α-trifluoro-2,6-dinitro-p-toluidine

Oryzalin controls annual grasses and broadleaf weeds in cotton, soybeans, nonbearing fruit trees, nut trees, vineyards, and ornamentals.

Pendimethalin and fluchloralin are among the more recent introductions to the nitroanaline herbicides. Pendimethalin controls annual grass and certain broadleaf weeds in field corn and preplant incorporated in cotton, soybeans, and tobacco. Fluchloralin is used for preplant and preemergence incorporated control of annual grasses and some broadleaf weeds in cotton and soybeans.

ORYZALIN (Surflan^R)

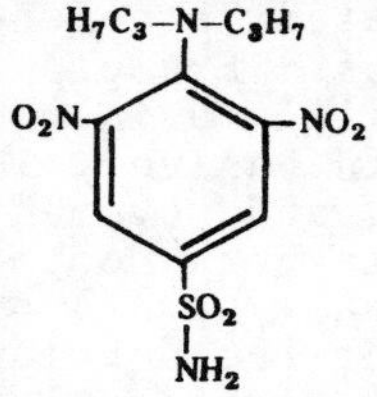

$H_7C_3-N-C_3H_7$

O_2N NO_2

SO_2

NH_2

3,5-dinitro-N^4,N^4-dipropylsulfanilamide

PENDIMETHALIN (Prowl^R)

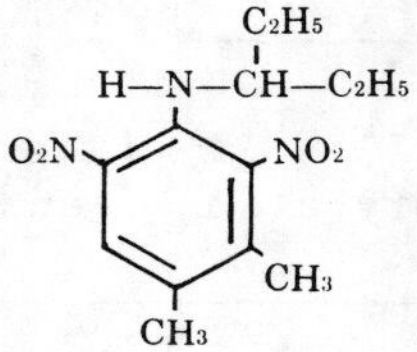

C_2H_5

$H-N-CH-C_2H_5$

O_2N NO_2

CH_3

CH_3

N-(1-ethylpropyl)-3,4-dimethyl-
2,6-dinitrobenzenamine

FLUCHLORALIN (Basalin^R)

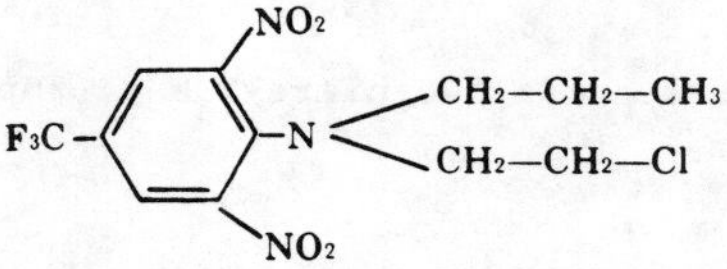

NO_2

$CH_2-CH_2-CH_3$

F_3C- N

CH_2-CH_2-Cl

NO_2

N-(2-chloroethyl)-a,a,a,-trifluoro-
2,6-dinitro-N-propyl-p-toluidine

   Ethalfluralin, structure not shown, is a new nitro-analine herbicide under experimental use permit for use in cotton and soybeans. Commercially it is known as Sonalan^R, and is used as a selective, soil-incorporated herbicide, and controls annual grasses and many broadleaf weeds. It is especially useful against nightshade.

6.24 Probably the group of herbicides most heavily used in
     agriculture is the _____.
- -

     nitroanilines

ISOPROPALIN

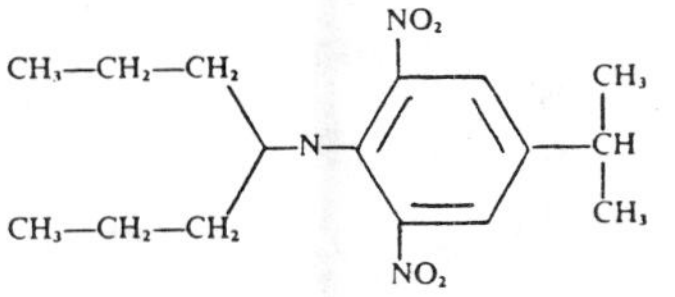

NITRALIN

CH₃—CH₂—CH₂

CH₃—CH₂—CH₂

N

NO₂

CH₃

CH

CH₃

NO₂

2,6-dinitro-N,N-dipropylcumidine

$$CH_3-\overset{\overset{O}{\|}}{\underset{\underset{O}{\|}}{S}}-\text{(ring)} NO_2, NO_2, -N(CH_2CH_2CH_3)_2$$

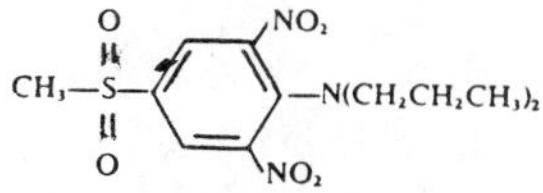

4-(methylsulfonyl)-2,6-dinitro-
N,N-dipropylaniline

6-G SUBSTITUTED UREAS

Urea is a simple nitrogen-containing organic molecule, $H_2N\overset{\overset{O}{\|}}{C}NH_2$. The substituted ureas have the hydrogen atoms replaced with various carbon chain and ring structures to yield a utilitarian group of compounds used primarily as selective preemergence herbicides. The ureas are strongly absorbed by the soil, then absorbed by roots. Their mechanism of action is inhibition of photosynthesis, (production of plant sugars), indirectly through inhibition of the Hill reaction. One other urea herbicide in heavy use is fenuron TCA (not illustrated).

6.25 The substituted ureas act by inhibiting _____.

- -

photosynthesis or Hill reaction

MONURON

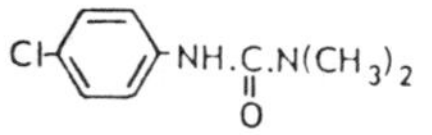

$$Cl-\text{(ring)}-NH.\overset{\overset{}{\underset{\underset{O}{\|}}{C}}}.N(CH_3)_2$$

3-(p-chlorophenyl)-1,1-dimethylurea

SIDURON (Tupersan[R])

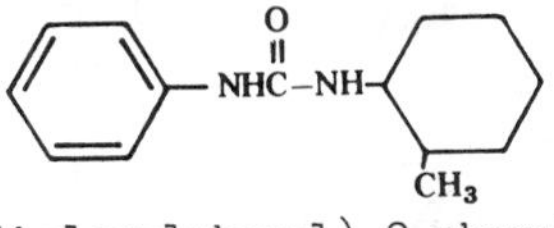

$$\text{(ring)}-NH\overset{\overset{O}{\|}}{C}-NH-\text{(cyclohexyl)} CH_3$$

1-(2-methylcyclohexyl)-3-phenylurea

DIURON (Karmex[R], Krovar[R])

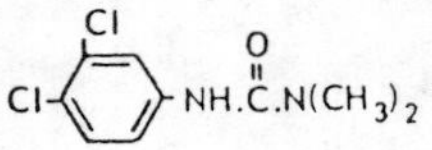

3-(3,4-dichlorophenyl)-1,1-dimethylurea

LINURON (Lorox[R])

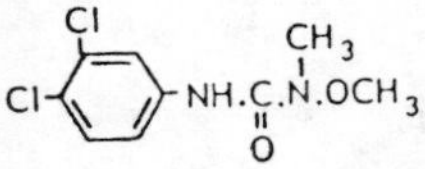

3-(3,4-dichlorophenyl)-1-methoxy-1-methylurea

Tebuthiuron is used for total vegetation woody plant control in noncropland areas, and brush and weed control in rangeland.

TEBUTHIURON (Spike[R])

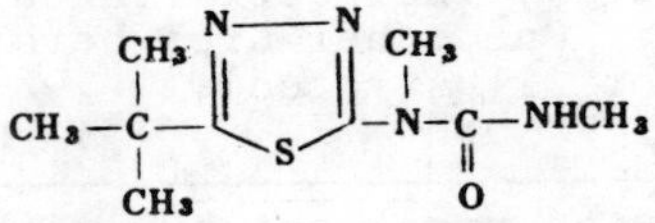

N-(5-(1,1-dimethyl)-1,3,4-thiadiazol
2-yl)-N,N'-dimethylurea

Fluometuron is a selective, preemergence, postemergence or layby treatment herbicide used exclusively on cotton. It is effective against several broadleaf and grass weeds, is not used against perennial weeds and may carry over in the soil with detrimental effects to next crop other than cotton.

FLUOMETURON (Cotoran[R])

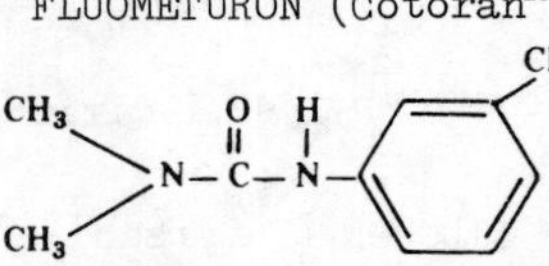

1,1-dimethyl-3-(a,a,a-trifluoro-m-tolyl)urea

6-H   CARBAMATES

As we have seen, some esters of carbamic acid

(HO-C-NH$_2$), the carbamates, are insecticidal, and, as we shall see later, others are fungicidal. Still other carbamates are herbicidal. Discovered in 1945, the carbamates are used primarily as selective preemergence herbicides, but some are also effective as postemergence ones.

The first of the carbamates was propham; it was followed first by chlorpropham, then by barban and terbucarb. These carbamates kill plants by stopping cell division and plant tissue growth. Two effects are noted: Cessation of protein production and shortening of chromosomes undergoing mitosis (duplication).

6.26 Carbamates kill weeds by stopping cell _____.

- -

        division

Chlorpropham has largely replaced propham for field crops, particularly in alfalfa, clovers, tomatoes, safflower and soybeans. Stored potatoes may also be treated to inhibit sprouting. Thiobencarb is a carbamate used as a postemergence, selective herbicide, for control of grasses and broadleaf weeds in rice fields.

PROPHAM (Chem-Hoe^R, IPC)

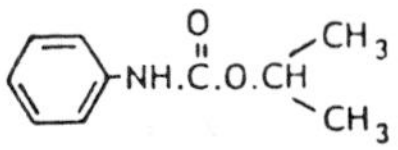

isopropyl carbanilate

CHLORPROPHAM (Chloro-IPC, Furloe^R)

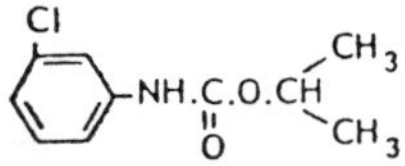

isopropyl m-chlorocarbanilate

BARBAN (Carbyne^R)

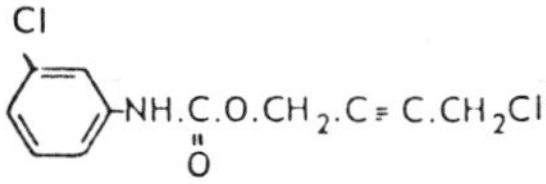

4-chloro-2-butynyl m-chlorocarbanilate

THIOBENCARB (Bolero^R, Saturn^R)

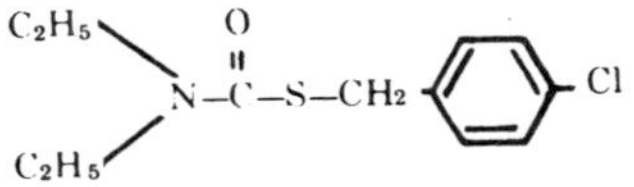

S-(4-chlorobenzyl)-N-N-diethylthiolcarbamate

Asulam is applied in most instances for grass weed
control. such as johnsongrass, crabgrass, foxtail, goose-
grass and barnyardgrass in sugarcane, but also for refores-
tration and Christmas tree plantings. Phenmedipham is used
entirely for broadleaf weeds in sugar beets and red table
beets.

ASULAM (Asulox^R)

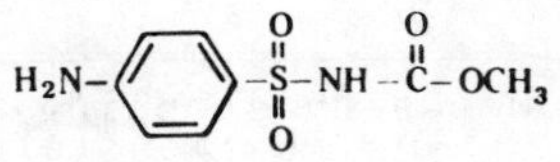

methyl sulfanilylcarbamate

PHENMEDIPHAM (Betanal^R)

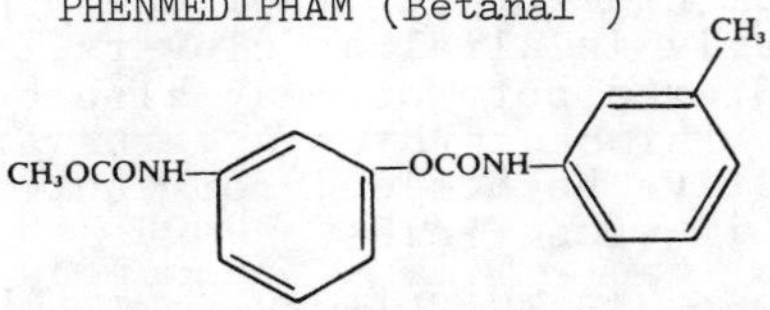

methyl m-hydroxycarbanilate m-methylcarbanilate

The 6 compounds just illustrated are aryl carbamates,
because they all contain an *aryl* group, the 6-member carbon *ring*, normally the phenyl ring; the *alkyl* group is the
carbon *chain* or hydrocarbon radical with one hydrogen displaced by attachment of the alkyl group to the remainder of
the molecule.

6.27 Aryl carbamates are those containing a _____
_____ in the molecule.

- -

phenyl ring

6.28 An alkyl group is the _____
or _____ _____ attached to a
molecule.

- -

carbon chain or hydrocarbon radical

6-I   THIOCARBAMATES

Within the carbamates are a group of thio- and dithiocarbamates which contain sulfur (from the Greek theion),
and are derived from the hypothetical thiocarbamic acid,
$$HS-\overset{\overset{O}{\|}}{C}-NH_2.$$ The thiocarbamates are selective herbicides
marketed for weed control in croplands. They are quite
volatile and must be incorporated in the soil after application. They inhibit the development of seedling shoots
and roots as they emerge from the weed seeds. Consequently, they are all used as either preplant or preemergence,
soil incorporated herbicides.

6.29 Thiocarbamates contain _____ appearing as
     the prefix thio- in the chemical name, taken from
     the Greek word for sulfur, _____.

- -

     sulfur   theion

6.30 The thiocarbamate herbicides have the same mechanism
     of action as the carbamates, inhibition of cell

     _____.

- -

     division

PEBULATE (PEBC, Tillam[R])   METHAM (Vapam[R])

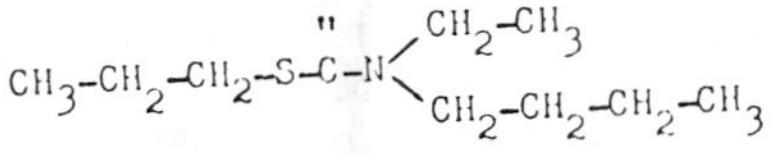

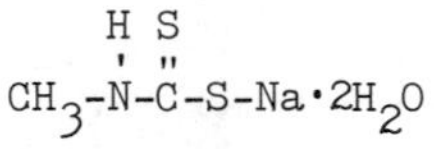

$$CH_3-CH_2-CH_2-S-\overset{\overset{O}{\|}}{C}-N\overset{\diagup CH_2-CH_3}{\diagdown CH_2-CH_2-CH_2-CH_3}$$

$$\overset{H\ \ S}{CH_3-N-\overset{\|}{C}-S-Na\cdot2H_2O}$$

S-propyl butylethylthiocarbamate   sodium N-methyldithiocarbamate

EPTC (Eptam[R])

$$C_2H_5S-CO-N(CH_2-CH_2-CH_3)_2$$

S-ethyl dipropylthiocarbamate

Metham is a dithiocarbamate, used as a general purpose soil fumigant for weed seeds, nematodes, insects and soil-borne disease microorganisms. In its metabolism in the soil it is converted to the isothiocyanate radical (-N=C=S), typical of the broad group of dithiocarbamate fungicides. This radical is highly toxic both to soil microorganisms and plant life.

Molinate is particularly effective for control of watergrass in rice. Cycloate is used in sugar beets and table beets to control annual and perennial grasses and many broadleaf weeds. Vernolate is effective for control of most of the grass weeds and some broadleaf weeds in soybeans and peanuts. Butylate is incorporated preplant to control most grass weeds in corn.

MOLINATE (Ordram[R])

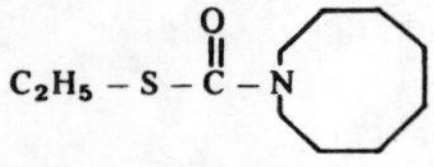

$$C_2H_5 - S - \overset{\overset{O}{\parallel}}{C} - N$$

S-ethyl hexahydro-1 H-azepine-1-carbothioate

CYCLOATE (Ro-Neet[R])

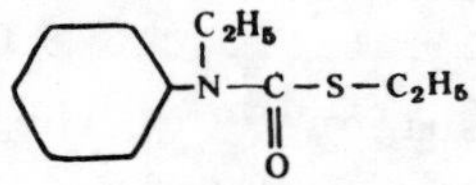

$$N - \overset{C_2H_5}{\underset{\underset{O}{\parallel}}{C}} - S - C_2H_5$$

S-ethylcyclohexylethylthiocarbamate

BUTYLATE (Sutan[R])

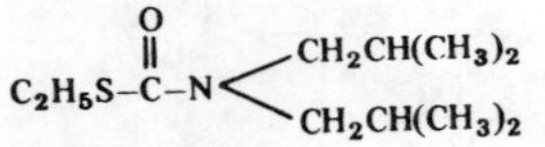

$$C_2H_5S - \overset{\overset{O}{\parallel}}{C} - N \begin{cases} CH_2CH(CH_3)_2 \\ CH_2CH(CH_3)_2 \end{cases}$$

S-ethyl diisobutylthiocarbamate

VERNOLATE (Vernam[R])

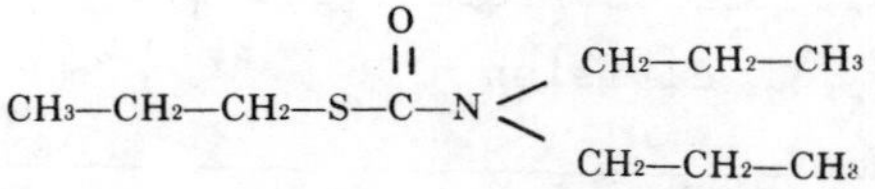

$$CH_3 - CH_2 - CH_2 - S - \overset{\overset{O}{\parallel}}{C} - N \begin{cases} CH_2 - CH_2 - CH_3 \\ CH_2 - CH_2 - CH_2 \end{cases}$$

S-propylidpropylthiocarbamate

6-J HETEROCYCLIC NITROGENS

The heterocyclic nitrogens are essentially triazines, which are 6-member rings containing 3 nitrogens (tri- = 3; and azine = a nitrogen-containing ring). The fundamental triazine nucleus is illustrated, showing the placement of the 3 nitrogens.

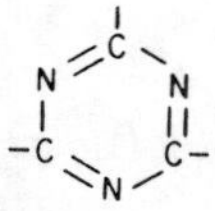

Triazine nucleus

6.31 Triazines are 6-member rings containing _____
_____.

- -

3 nitrogens

The triazines are strong inhibitors of photosynthesis, and their selectivity depends on the ability of tolerant plants to degrade or metabolize the parent compound whereas the susceptible plants do not. Triazines are applied to the soil primarily for their post-emergence activity. They are used in greatest quantity in corn production and non-selectively on industrial sites.

6.32 Triazines achieve selectivity by affecting those
plants which cannot _____ the herbicide,
through the inhibition of _____.

- -

metabolize   photosynthesis

There are many triazines on the market today, several of which are illustrated below. Cyanazine is used on corn, sorghum and cotton, while prometon is for noncrop weed control only. Atrazine is the oldest of the triazines and used in greatest quantity on corn.

ATRAZINE

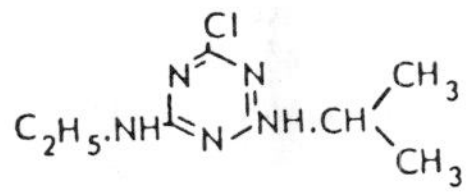

$C_2H_5.NH$ ... $NH.CH$ CH_3 / CH_3 (Cl)

2-chloro-4-(ethylamino)-6-

(isopropylamino)-s-triazine

CYANAZINE (Bladex^R)

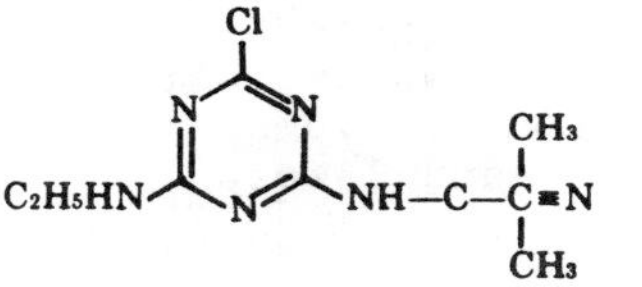

C_2H_5HN ... $NH-C-C\equiv N$ CH_3 CH_3 (Cl)

2-(4-chloro-6-ethylamino-s-triazin-2-ylamino)-2-methylpropionitrile

PROMETON (Pramitol^R)

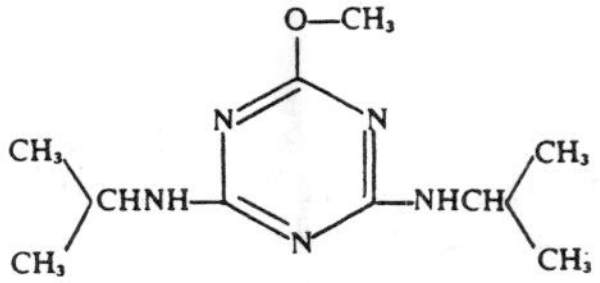

CH_3 CH_3 $CHNH$ ... $NHCH$ CH_3 CH_3 (O—CH₃)

2,4-bis(isopropylamino)-6-
methoxy-s-triazine

SIMAZINE

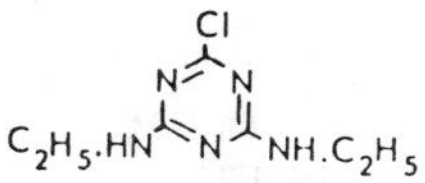

$C_2H_5.HN$ ... $NH.C_2H_5$ (Cl)

2-chloro-4,6-bis(ethylamino)-

s-triazine

Metribuzin is used on soybeans, wheat, sugarcane and a few vegetables. Hexazinone controls annual, biennial and perennial weeds and woody plants on noncropland. Sometimes applied as pellets or gridballs to control woody plants in conifer plantings.

METRIBUZIN (Lexone[R], Sencor[R])

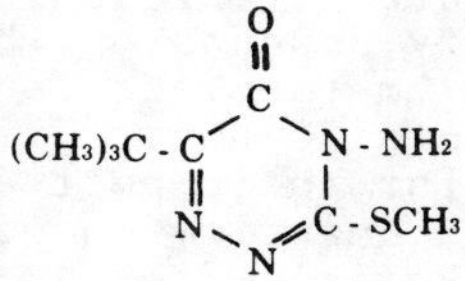

4-amino-6-(1,1-dimethylethyl)-3-
(methylthio)-1,2,4-triazin-5(4H)-one

HEXAZINONE (Velpar[R])

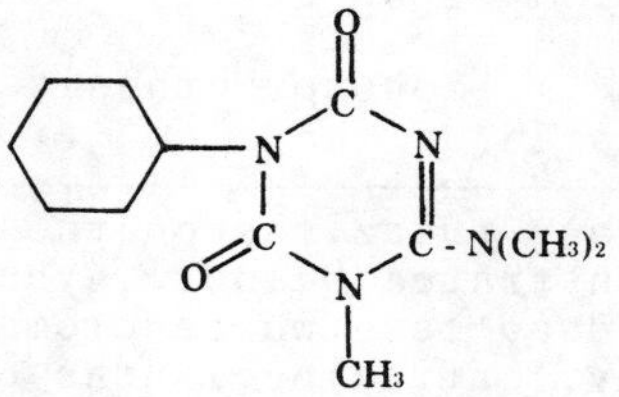

3-cyclohexyl-6-(dimethylamino)-1-
methyl-1,3,5-triazine-2,4-(1H,3H)-dione

Chlorsulfuron is used as a selective, preemergence and postemergence herbicide. It is particularly useful on small grains, wheat, barley and oats, at the phenominally low rates of 1/6 to 1.0 ounce per acre for the control of many difficult annual weeds.

CHLORSULFURON (Glean[R])

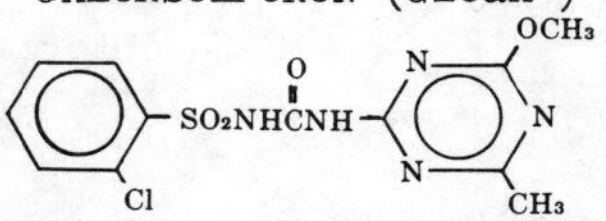

2-chloro-N-((4-methoxy-6-methyl-1,3,5-triazin-2-yl)
aminocarbonyl) benzenesulfonamide

Triazoles are 5-member rings containing 3 nitrogens. The outstanding member of this group is amitrole, the notorious aminotriazol which became embroiled in the historic "cranberry incident" of 1959. It was this event that culminated in the addition of the Delaney Amendment in the Pure Food and Drug Act in 1960 (See Unit 13, Pesticides and the Law). In essence, this Amendment stated that no residue

of any cancer-producing agent (carcinogen) would be toler-
ated in a food crop. But more about that later. Amitrole
acts in about the same way as the triazines, by inhibiting
photosynthesis.

AMITROLE

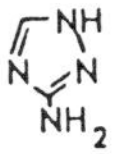

3-amino-s-triazole

Oxadiazon is used for weed control in dry-seeded rice,
and for turf and ornamentals. Bentazon is effective in
controlling broadleaf weeds in soybeans, rice, corn, peanuts, beans and peas.

OXADIAZON (Ronstar^R)   BENTAZON (Basagran^R)

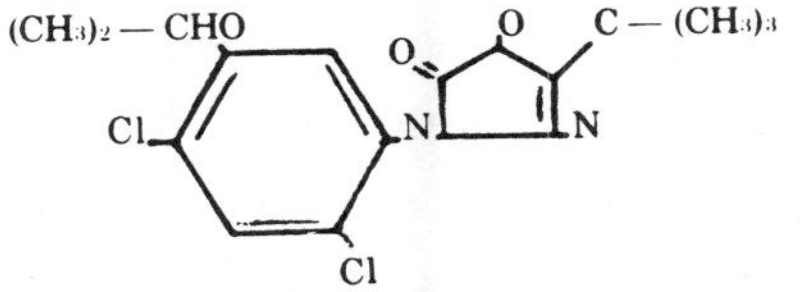

2-tert-butyl-4-(2,4-dichloro-
5-isopropoxyphenyl)-Δ^2-1,3,4-
oxadiazolin-5-one

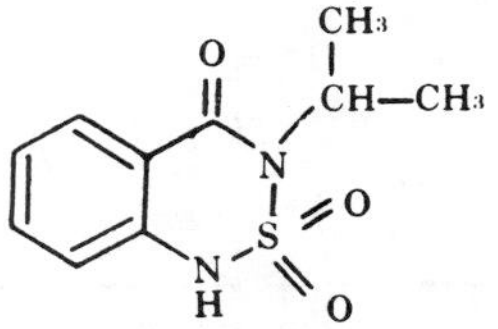

3-(1-methylethyl)-1H-2,1,3-benzo-
thiadiazin-4(3H)-one 2,2-dioxide

Pyrazoles are 5-member rings containing 2 nitrogens.
The only herbicide in this group for the moment is difenzoquat (Avenge^R). Difenzoquat is highly selective against
wild oats in grain crops. It has no preemergence activity
but is applied post emergence when wild oats are in the 3-5
leaf stage.

DIFENZOQUAT (Avenge^R)

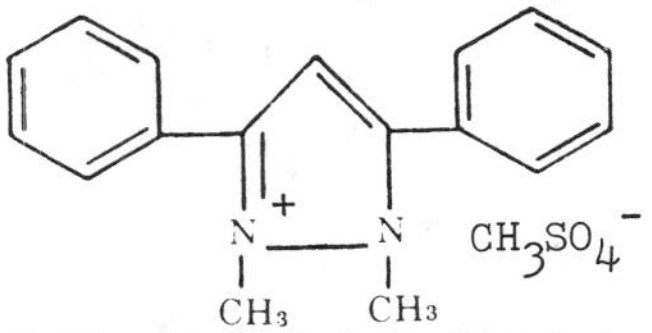

1,2-dimethyl-3,5-diphenyl-1
H-pyrazolium methyl sulfate

Picloram is a popular herbicide derived from the pyridine molecule, both of which are illustrated. Picloram is a readily translocated herbicide used against broadleaved and woody plants; it may be taken up from either the roots or the foliage. It has also been used experimentally as a growth regulator at extremely low dosages on apricots, cherries, and figs. Its mechanism of action is probably the regulation of protein and enzyme synthesis in cells through its effect on nucleic acid synthesis and metabolism.

PYRIDINE

PICLORAM (Grazon[R], Tordon[R])

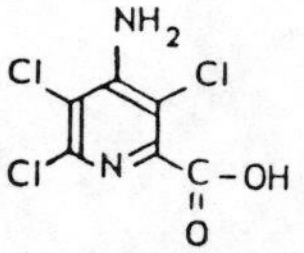

4-amino-3,5,6-trichloropicolinic acid

6.33 Picloram affects the synthesis of _____
     and _____.

- -

protein and enzymes

A relatively new brush-control compound is triclopyr, closely related to picloram. It is used as a selective, translocated, postemergence herbicide, and controls all types of brush in rights-of-way, industrial and forestry sites, and experimentally for rangelands.

TRICLOPYR (Garlon[R])

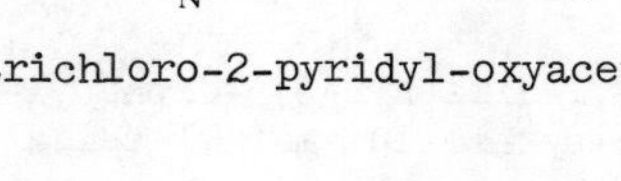

3,5,6-trichloro-2-pyridyl-oxyacetic acid

Sulfometuron-methyl is a broad spectrum, pre-and post-emergence herbicide, used for nonselective weed control in noncrop areas.

SULFOMETURON-METHYL (Oust^R)

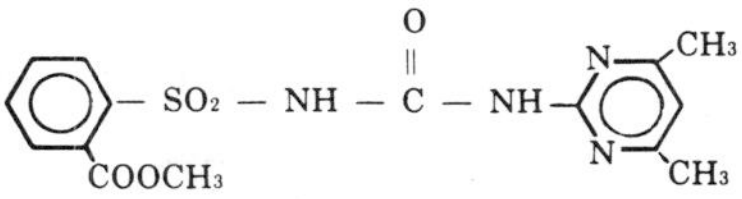

methyl 2-(((((4,6-dimethyl-2-pyrimidinyl) amino)-carbonyl)amino)sulfonyl)benzoate

Scepter^R is new enough not to have a common name assigned. It is an imidazole compound used as a selective, pre-and postemergence herbicide. At this writing it holds a Section 18 (emergency registration) for the control of sicklepod in soybeans in Alabama, Louisiana and Mississippi. Most broadleaf weeds are susceptible to its relatively slow action.

Scepter^{R}H (AC 252,214)

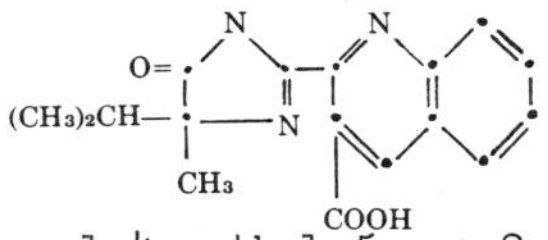

2-(4-isopropyl-4-methyl-5-oxo-2-imidazolin-2-yl)-3-quinolinecarboxylic acid

The substituted uracils give a wide range of grass and broadleaf control for an extended period by inhibition of photosynthesis. These include bromacil, terbacil and lenacil (Venzar^R) (not illustrated).

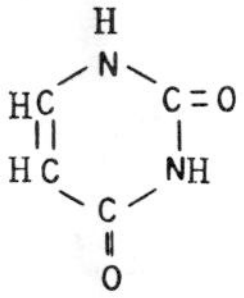

Uracil nucleus

BROMACIL (Hyvar^R)

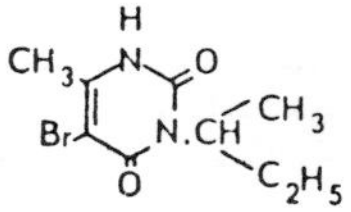

5-bromo-3-sec-butyl-6-methyluracil

TERBACIL (Sinbar[R])

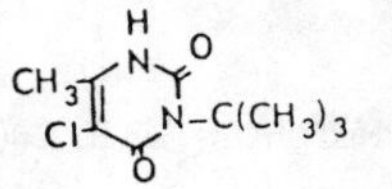

$$CH_3 \quad \overset{H}{\underset{Cl}{N}} \quad \overset{O}{\underset{N-C(CH_3)_3}{}}$$

3-<u>tert</u>-butyl-5-chloro-6-methyluracil

6.34 The substituted uracils control weeds by

----

inhibiting photosynthesis

6-K <u>ALIPHATIC ACIDS</u>

An aliphatic acid is a carbon chain acid. Two heavily used herbicides in this group are TCA and dalapon, used against grasses, particularly our old enemies (or friends, depending on your view), quackgrass and Bermudagrass. They both act by precipitation of protein within the cells.

TCA

$$\underset{Cl}{\overset{Cl}{ClC}} - \overset{O}{C} - OH$$

trichloroacetic acid

DALAPON (Dowpon[R])

$$CH_3 - \underset{Cl}{\overset{Cl}{C}} - \overset{O}{C} - OH$$

2,2-dichloropropionic acid

6.35 TCA and dalapon are members of the _____
                                    group, and kill grasses by
_____   of _____.

----

aliphatic acid   precipitation of protein

6-L ARYLALIPHATIC ACIDS (SUBSTITUTED BENZOIC ACIDS)

The aliphatic acids are carbon chain acids. Thus, the arylaliphatic acids must be aryls, or 6-member rings, attached to aliphatic acids. A number of these materials are employed as herbicides, and are applied to the soil against germinating seeds and seedlings. The mechanism of action for dicamba and fenac is not completely understood; however, it is probably similar to that of the phenoxy herbicides (2,4-D, etc.), which interfere with the nucleic acid metabolism.

DICAMBA (Banvel^R, Banex^R)   FENAC

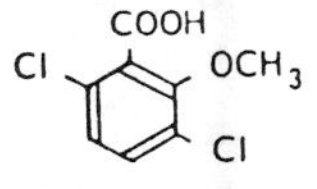

2-methoxy-3,6-dichlorobenzoic acid   (2,3,6-trichlorophenyl)acetic acid

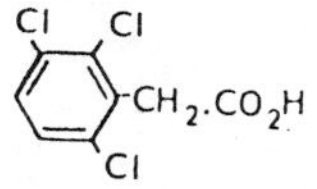

The growth effects of DCPA and chloramben are auxin-like, but the available research data are insufficient to propose a mechanism of action

DCPA (Dacthal^R)   CHLORAMBEN (Amiben^R)

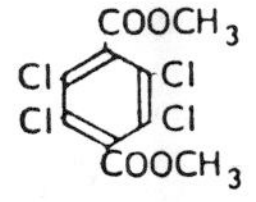

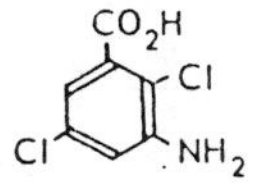

dimethyl tetrachloroterephthalate   3-amino-2,5-dichlorobenzoic acid

---

5.36 Several of the effects of arylaliphatic acid herbicides on plants resemble those of phenoxy compounds, indicating that they probably interfere in the metabolism of _____ _____.

- -

nucleic acids

6-M PHENOL DERIVATIVES

The phenol derivatives are highly toxic, nonselective foliar herbicides which are most effective in hot weather. The nitrophenols, represented by DNOC, were first introduced as herbicides in 1932. Dinoseb is the best known representative of the nitrophenols, which act by uncoupling oxidative phosphorylation. Though rather old, the phenols are still used in great quantity as general contact herbicides, usually against broadleaf weeds, and selectively in cereal crops.

The dinitrophenols are familiar compounds, since they have already been discussed under the insecticides. There, you will remember, one or more members of this group have also been used as ovicides, insecticides, fungicides, and blossom-thinning agents.

6.37 Nitrophenols not only kill plants but all living
     material at some dosage by _____

     _____   _____.

- -

     uncoupling oxidative phosphorylation

6.38 What one word would you use to describe a pesticide
     which is apparently lethal to all living material?

     _____

- -

     biocide

DNOC   DINOSEB, DINITRO, DNBP

O_2N—(ring)—CH_3, OH, NO_2 $CH_3.CH_2.CH$—(ring), CH_3 OH, NO_2, NO_2

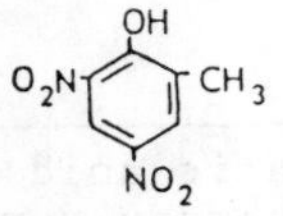

4,6-dinitro-o-cresol   2-sec-butyl-4,6-dinitrophenol

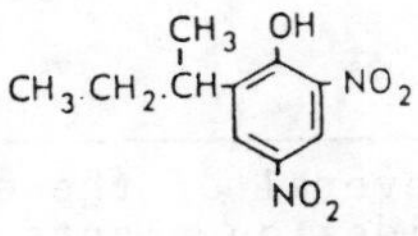

A second sub-group of the phenols is the chlorinated phenols, having only one member recognized as a herbicide, PCP or pentachlorophenol. Because of its broad action, it is also used for termite protection and for wood treatment to prevent fungal rots. In the past it was used as a nonselective herbicide, and as a pre-harvest defoliant. Its mechanism of action is a combination of plasmolysis, protein precipitation, and desiccation. Because of its wide effectiveness and multiple routes of action, PCP is recognized as being destructive to all living cells. It is a restricted use material, registered only as a wood preservative and for the control of subterranean termites.

PCP, PENTA

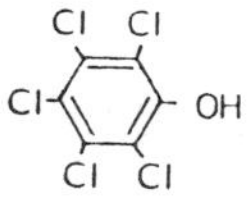

pentachlorophenol

6.39 Based on its mechanism of action, what would be the effect of PCP on the skin? _____

- -

highly irritating

6-N SUBSTITUTED NITRILES

A nitrile is any organic compound containing the C≡N or cyanide grouping. There are several substituted nitrile herbicides, which have a wide spectrum of uses against grasses and broadleaf weeds. Their mechanisms of action are broad, involving seedling growth inhibition, potato sprout inhibition, and gross disruption of tissues through inhibition of oxidative phosphorylation and preventing the fixation of CO_2. These effects, of course, do not explain their rapid action. This is attributed to rapid permeation. Two popular members of the nitriles, dichlobenil and bromoxynil, are illustrated. ME4 Brominal[R], the organic acid ester formulation of bromoxynil, was registered for application through sprinkler irrigation systems in 1985. This method of application is currently referred to as chemigation, from the application of pesticidal chemicals through the irrigation system, chemical-irrigation or chemigation.

DICHLOBENIL (Casoron[R])   BROMOXYNIL (Brominal[R], Buctril[R])

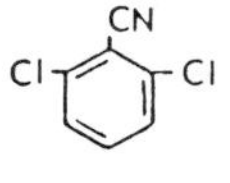

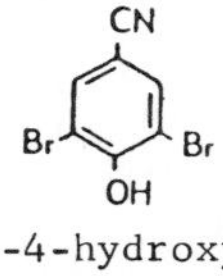

2,6-dichlorobenzonitrile   3,5-dibromo-4-hydroxybenzonitrile

6.40 The substituted nitriles inhibit _____
                              and prevent fixation of _____
     _____ .

- -

     oxidative phosphorylation CO_2

6-O BIPYRIDYLIUMS

     The name bipyridylium, suggests the attachment of two
pyridyl rings. Beyond this it is necessary to know only
that there are two important herbicides in this group,
diquat and paraquat. Both are contact herbicides that
damage plant tissues quickly, causing the plants to appear
frostbitten because of cell membrane destruction. This
wilting and desiccation occur within hours making these
novel herbicides also useful as preharvest desiccants for
seed crops, cotton, soybeans, sugarcane and sunflowers.
Diquat is also used in aquatic weed control. Both materials are deactivated almost as soon as they come in contact with soil.

6.41 Paraquat and diquat cause plants to appear _____
                        because they _____the cell
     membranes.

- -

     frostbitten   rupture or destroy

DIQUAT

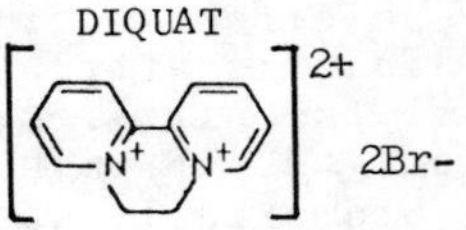

6,7-dihydrodipyridol(1,2-a:2',1'-c)pyrazidinium
                (dibromide)

PARAQUAT

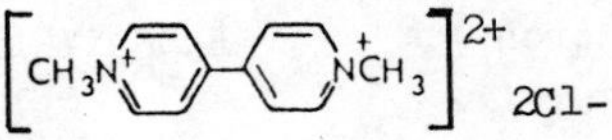

I,1'-dimethyl-4,4'-bipyridylium ion
            (dichloride)

6-P <u>CINEOLES</u>

The newest chemical class of herbicides is the cineoles, of which there is only one member, cinmethylin (SD 95481), introduced in 1985 under experimental use permits. It is a preemergence or incorporated herbicide useful for the control of many annual grasses and certain broadleaved weeds in cotton, soybeans, and peanuts. Other crops are under investigation. Its mode of action is that of disrupting development of meristematic (growing) tissues in shoots and roots of susceptible species. It is not effective against older, more mature weeds.

Notice that the chemical structure contains only carbon, hydrogen and oxygen atoms, an unusual characteristic for any synthetic pesticide.

CINMETHYLIN (Cinch[R])

7-oxabicyclo(2.2.1)heptane, 1-methyl-4-
-(1-methylethyl)-2-((2-methylphenyl)-
methoxy)-, exo-

6-Q <u>MICROBIALS</u> <u>OR</u> <u>MYCO-HERBICIDES</u>

A new concept in weed control is the use of disease microorganisms as useful, sometimes self-perpetuating pathogens. The first to be registered in this field is *Phytophthora palmivora* (DeVine[R]), a naturally occurring, highly selective disease of milkweed vine (*Morrenia odorata*), a serious pest in citrus groves. When properly applied to the soil beneath citrus trees, it kills existing milkweed vines. The microorganism persists in the vine root debris and continues to control germinating vines for more than one year after a single treatment. This pathogen is selective and does not infest citrus roots, however because certain other ornamental crops are susceptible, caution is recommended.

The most recent myco-herbicide is Collego[R], the spores of *Colletotrichum gloeosporioides*. It is currently registered for use in rice and soybeans for the control of jointvetch or curly indigo. The spores are sensitive to excessive temperatures in storage or in the spray tank, as well as to all synthetic pesticides remaining in the tank from previous use.

6-R MISCELLANEOUS HERBICIDES

Among the miscellaneous herbicides belong methyl bromide (CH_3Br) which is used as a fumigant for any known organism in soil: nematodes, fungi, seed, insects, and other plant parts. Allyl alcohol ($CH_2=CHCH_2OH$) is a volatile, water-soluble fumigant used for the same purposes as methyl bromide.

Endothall is used as an aquatic weed and a selective field crop herbicide. It acts by interfering in RNA synthesis. Endothall has one distinct advantage over most aquatic weed killers, its low toxicity to fish, an outstanding example of environmental protection through pesticide selectivity.

ENDOTHALL

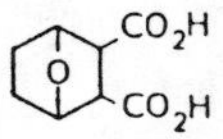

7-oxabicyclo(2,2,1)heptane-2,3-dicarboxylic acid

6.42 What word best describes the action of a herbicide
     which is effective against one species but not
     another?

- -

selective

Glyphosate (Roundup[R]) was discovered in 1971 and belongs to the glyphosphate herbicide classification. It is a non-selective, nonresidual, postemergence material. Glyphosate is recognized for its effectiveness against perennial, deep-rooted, grass and broadleaf weeds, as well as woody bush problems in crop and noncrop areas. It is a translocated, foliar applied herbicide that can be applied at any stage of plant growth or time of year, with most types of application equipment, including the new wick, roller, and wiper devices. Its mechanism of action appears to be the inhibition of nucleic acid metabolism and protein synthesis, by way of aromatic acid synthesis inhibition.

GLYPHOSATE (Roundup[R])

N-(phosphonomethyl) glycine
(isopropylamine salt)

Sethoxydim is a relatively new, selective, postemergence material, highly selective for annual and perennial
grasses. It is currently registered for soybeans, cotton
and ornamentals, and is being tested on most broadleaf crops
including vegetables.

SETHOXYDIM (Poast^R)

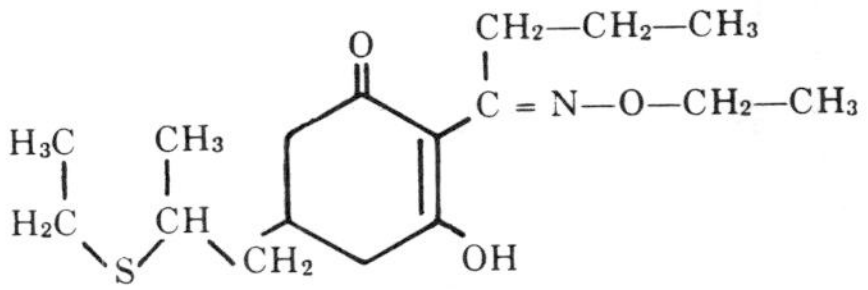

2-(1-(ethoxyimino)butyl)-5-(2-(ethylthio)
propyl)-3-hydroxy-2-cyclohexene-1-one

Bensulide is one of the better turf herbicides, expe-
cially for the control of crabgrass. It is an organophosphate, but is considered one of the less toxic herbicidal
materials. Bensulide acts by inhibiting cell division in
root tips. It is used as a preemergence herbicide in lawns
to control certain grasses and broadleaf weeds, but it fails
as a foliar spray because it is not translocated.

BENSULIDE (Betasan^R, Prefar^R)

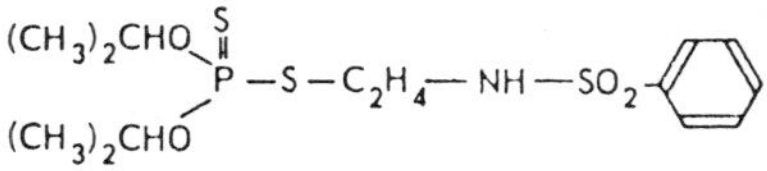

O,O-diisopropyl phosphorodithioate S-ester with

N-(2-mercaptoethyl)benzenesulfonamide

6.43 Bensulide inhibits _____ _____
     in root tips.
- -

     cell division

Acrolein is an extremely useful aquatic weed herbicide,
but it requires *application only by certified and licensed
operators* because of its frightening tear-gas effect. Plants
exposed to acrolein disintegrate within a few hours and
float downstream. It is a general plant toxicant, destroying plant cell membranes and reacting with various enzyme
systems. By using spot treatment in lakes, the fish population can be saved. Another example of environmental protection, but by a different method, spot application.

ACROLEIN

$$CH_2=CH-CHO$$

Fosamine ammonium belongs to the organophosphate herbicides and is used as a foliar spray for brush control on non-cropland, principally rights-of-way.

FOSAMINE AMMONIUM (Krenite[R])

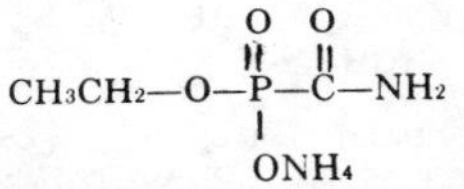

$$CH_3CH_2-O-\overset{\overset{O}{\|}}{P}-\overset{\overset{O}{\|}}{C}-NH_2$$
$$\underset{ONH_4}{|}$$

ammonium ethyl carbamoylphosphonate

Dimethazone (Command[R]) is the first of a new group of herbicides, the isoxazolidinones. Dimethazone, applied either preemergence or preplant incorporated, controls many broadleaf and grass weeds in soybeans. It is absorbed by roots and shoots and translocated throughout the leaf. Its mode of action is the inhibition of chlorophyll- and caro-tenoid-synthesis. It acts by inhibiting the synthesis of chlorophyll and carotenoids. As a result affected plants emerge from the soil devoid of pigmentation and quickly die.

DIMETHAZONE (Command[R])

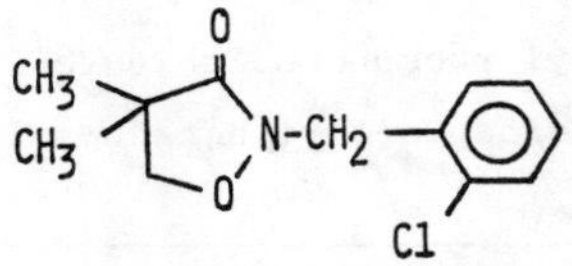

2-(2-chlorophenyl)methyl-4-
dimethyl-3-isoxazolidinone

Summary: There are many chemical and use classes of herbicides available, some of which have different mechanisms of action within the same chemical class. Those discussed and illustrated here are but a cross-section of the existing herbicides. We can expect to see different classes develop in the future, with mechanisms of action more clearly delin-eated, following intensive research on this complex subject. Because of the wealth of materials and action, no good sum-mary can be made. Rather, the reader is urged to review the groups frequently, to establish generalizations regarding herbicide use and action classes.

6.44 For a herbicide review, try identifying the structures below as to their classification:
inorganics____; arsenicals ____; phenoxyaliphatic acids____; substituted amides____; nitroanilines____; substituted ureas____; carbamates____; heterocyclic nitrogens____; aliphatic acids____; arylaliphatic acids____; phenol derivatives____; substituted nitriles____; bipyridyliums____; diphenyl ethers____.

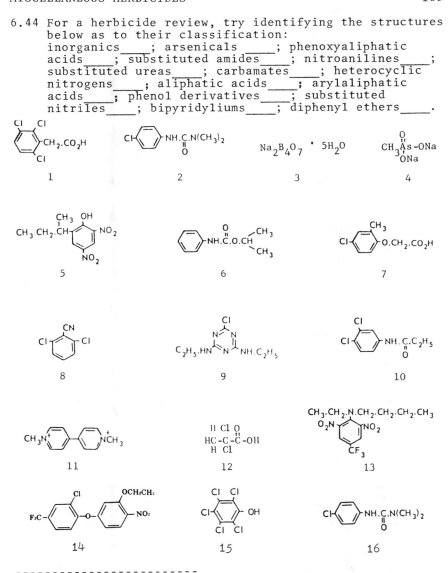

inorganics__3__; arsenicals__4__; phenoxyaliphatic acids__7__; substituted amides__10__; nitroanilines__13__; substituted ureas__2__; carbamates__6__; heterocyclic nitrogens__9__; aliphatic acids__12__; arylaliphatic acids__1__; phenol derivatives__5__; substituted nitriles__8__; bipyridyliums__11__; diphenyl ethers__14__.

FUNGICIDES AND BACTERICIDES

Most plant diseases can be controlled to some extent with existing fungicides. Among those that previously could not, but are beginning to be controlled with chemicals are the root rots caused by *Phytophthora* and *Rhizoctania* and *Fusarium*, *Verticillium* and bacterial wilts. These diseases occur either below ground, and beyond the reach of traditional fungicides, or are systemic within the plant.

Though fungicides, strictly speaking, are chemicals used to kill fungi, we shall consider them as chemicals used to control bacterial as well as fungal plant pathogens, the causal agents of many plant diseases. Other plant disease organisms are viruses, mycoplasma-like organisms, rickettsias, some insects, algae, and parasitic seed plants. For the purposes of our discussion, we shall deal only with chemicals applied for the control of fungi and bacteria.

7.1 Plant diseases are caused by _____,
_____, _____,
_____, or _____.

- -

fungi, bacteria, viruses, nematodes, parasitic plants, mycoplasma-like organisms, rickettsias, some insects, algae

---

Examples of plant diseases are legion. They include storage rots, seedling diseases, root rots, gall disease, vascular wilts, leaf blights, rusts, smuts, mildews, and viral diseases. These can, in many instances, be controlled by the early and continued application of selected fungicides that either kill the pathogenic organisms or inhibit their development.

Fungal diseases are basically more difficult to control with chemicals than insects because the fungus itself is a plant living in close quarters with its host. This explains the difficulty in finding chemicals which kill the fungi without harming the plant. Also, fungi that can be controlled by fungicides may undergo secondary cycles rapidly and thus produce 12-25 "generations" during a

3-month crop-growing season. Consequently, repeated applications of protective fungicides may be necessary, due to
plant growth dilution, removal by rain and the usual physical and chemical degradation.

7.2 Why are some plant diseases extremely difficult to
     control?

     They are below ground and beyond reach, or are
     systemic.

7.3 Why are fungal diseases of plants in general difficult
     to control?
     A _____
     B _____

     A. Fungi are also plants
     B. Fungi reproduce rapidly, many generations/season

     Our present arsenal contains about 140 fungicidal
materials, most of which are recently discovered organic
compounds. Most of these act as *protectants*, preventing
spore germination and subsequent fungal penetration of
plant tissues. Protectants are applied repeatedly to
cover new plant growth and to replenish the fungicide that
has deteriorated or has been washed off by rain.

7.4 Most fungicides are placed on the plant as _____
     before the disease strikes.

---

     protectants

     Thanks to modern chemistry, many of the serious
diseases of grain crops are controlled by seed treatment
with selective materials. Others are controlled with
resistant varieties. Diseases of fruit and vegetables are
often controlled by sprays or dusts of fungicides.
     Historically, fungicides have centered around sulfur,
copper, and mercury compounds, and, even today, most of
our plant diseases could be controlled by these groups.
However, the sulfur and copper compounds can retard
growth in sensitive plants, and, as a result, the organic
fungicides were developed. These sometimes have greater
fungicidal activity and usually have less *phytotoxicity*.

7.5   The inorganic sulfur and copper fungicides are some-
      times _____ to the more sensitive
      plants, and have been replaced by the modern
      _____ fungicides.
- -

      phytotoxic   organic

Fungicides should be applied to protect plants during
stages when they are vulnerable to inoculation by pathogens, before there is any evidence of disease. Fungicides
can help to control certain diseases *after* the symptoms
appear, referred to as *chemotherapeutants*. Also, protective fungicides are commonly used, even after symptoms of
disease have appeared. *Eradicant* fungicides are usually
applied directly to the pathogen during its "over-winter-
ing" stage, long *before* disease has begun and symptoms have
appeared, However, if sale of the crop depends on its
appearance, such as lettuce and celery, then the fungicide
must be applied as a protectant spray in advance of the
pathogens to prevent the disease.

7.6   Fungicides are usually applied to plants when they
      are _____ to the pathogens as a _____.
      However, after the disease becomes established, it
      may be controlled with a _____.
- -

      susceptible   protectant   chemotherapeutant

7.7   _____ are applied long before the
      disease strikes, while it overwinters.
- -

      eradicants

The principle of fungicidal application differs from
the application of herbicides and insecticides. Only that
portion of the plant which has a coating of dust or spray
film of fungicide is protected from disease. Thus good uniform coverage is essential. Fungicides, with several new
exceptions, are not systemic in their action. They are
applied as sprays or dusts, but sprays are preferable since
the films stick more readily, remain longer, can be applied
most any time of the day, and result in less off-target
drift.
      The general-purpose fungicides include inorganic forms
of copper, sulfur, and metallic complexes of cadmium,

chromium, and zinc, along with a variety of organic compounds. The general-use fungicides for lawn and garden are
few and are usually organic compounds.
    We have arrived at the fungicides themselves, which
can best be examined historically. The inorganic fungicides were first on the scene.

INORGANIC FUNGICIDES

7-A SULFUR

    Elemental sulfur in many forms is probably the oldest
fungicide known. There are three physical forms or formulations of sulfur used as fungicides. The first is finely
ground sulfur dust which contains 1% to 5% clay or talc to
assist in dusting qualities. The sulfur in this form may
be used as a carrier for another fungicide or an insecticide. The second is flotation or colloidal sulfur, which
is so very fine that it must be formulated as a wet paste
in order to be mixed with water. In its original dry
microparticle size, it would be impossible to mix with
water, rather it would merely float. Wettable sulfur is the
third form, which is finely ground with a wetting agent so
that it will mix readily with water for spraying. For most
effective disease control the particle size of wettable
sulfurs should be no larger than 7 μm. Dusting sulfur
should pass through a 325-mesh or finer screen.

7.8 The three formulations of elemental sulfur are
    _____ , _____ , and _____ .

- -

     dust   paste   wettable powder

    Sulfur kills certain pathogenic organisms (and mites)
by direct contact, and also by fumigant action at temperatures above 70°F. The fumigant effect is, however, somewhat secondary at marginal temperatures and under windy conditions. At temperatures above 90°F sulfur is frequently
phytotoxic! It is quite effective in controlling powdery
mildews of plants that are not unduly sensitive to sulfur.
Unlike those of any other fungus, spores of powdery mildews
will germinate in the absence of a film of water in the penetration court. Its fumigant effect - "acting at a distance" - is undoubtedly important in killing spores of powdery mildews. It is absorbed by fungi in the vapor state,
thus its fumigant action. Sulfur interferes in electron
transport along the cytochromes, and is then reduced to
hydrogen sulfide (H_2S), a toxic entity to most cellular
proteins.

7-B COPPER

Most of the inorganic copper compounds are practically
insoluble in water, and are likely to be the pretty blue,
green, red, or yellow powders sold as fungicides. The
various forms include Bordeaux mixture, named after the
Bordeaux region in France where it originated. Bordeaux
is a chemically undefined mixture of copper sulfate and
hydrated lime, which was accidentally discovered when
sprayed on grapes in Bordeaux to scare off "freeloaders."
It was soon observed that a disease of grapes, downy mildew, disappeared from the treated plants. From this unique
origin began the commercialization of fungicides. The
copper ion, which becomes available from both the highly
soluble and relatively insoluble copper salts, provides the
fungicidal as well as phytotoxic and poisonous properties.
A few of the many inorganic copper compounds used over the
years are presented in Panel G.
    A comment on solubility is appropriate at this point.
In general, protective fungicides have low ionization con-
stants, but, in water, some toxicant does go into solution.
The small quantity absorbed by the fungal spore is then
replaced in solution from the residue. The spore accumulates the toxic ion (or radical) and, so to speak, "commits
suicide." Except for powdery mildews, water in the penetration court permits spore germination and solubilizes the
toxic portion of the fungicidal residue.
    The copper ion is toxic to all plant cells, and must
be used in discrete dosages or in relatively insoluble
forms, to prevent phytotoxicity to the host plant. This
is the basis for the use of relatively insoluble or
"fixed" copper fungicides.

7.9 How can relatively insoluble copper salts kill
     fungal spores?

Spores continue removing copper ions from solution —
suicide.

Copper compounds are not easily washed from leaves
by rain since they are relatively insoluble in water;
they give longer protection against disease than do most
of the organic materials. They are relatively safe to
use and require no special precautions during spraying.
Despite the fact that copper is an essential element for
plants, there is some danger in a buildup of copper in
the soil resulting from frequent and prolonged use. In
fact certain citrus growers in Florida had a serious problem of copper toxicity after using fixed copper for disease
control.

PANEL G

Some of the vast number of inorganic copper compounds
used as fungicides.

Name	Chemical Formula	Uses
Cupric sulfate	$CuSO_4.5H_2O$	Seed treatment and preparation of Bordeaux mixture
Cupric hydroxide (Kocide[R])	$Cu(OH)_2$	Seed treatment, foliage spray; many fungal diseases
Copper oxychloride	$3Cu(OH)_2.CuCl_2$	Powdery mildews
Copper oxychloride sulfate	$3Cu(OH)_2.CuCl_2$ + $3Cu(OH)_2.CuSO_4$	Many fungal diseases
Copper ammonium carbonate (Copper-Count-N[R])	Chemical complex (formula not known)	Many citrus, deciduous fruit and vegetable diseases
Cuprous oxide	Cu_2O	Powdery mildews
Basic copper sulfate	$CuSO_4.Cu(OH)_2.H_2O$	Seed treatment and preparation of Bordeaux mixture
Cupric carbonate (Malachite[R])	$Cu(OH)_2.CuCO_3$	Many fungal diseases
Copper resinate* (Citcop[R])	Salts of fatty & rosin acids	Bacterial & fungal diseases of grapes, citrus, vegetables

* This is actually an organic salt of copper.

The currently accepted theory for the mode of action
of copper's fungistatic action is its nonspecific denatur-
ation of protein. The Cu^{++} ion reacts with enzymes having
reactive sulfhydryl groups which would explain its toxicity
to all forms of plant life, but especially its toxicity to
the vulnerable copper-concentrating spores and cells.

7.10 Copper's fungistatic action is attributed to its
     nonspecific denaturation of _____ and
     reactivity with enzymes having _____ groups.

- -

     protein   sulfhydryl

7-C <u>MERCURY</u>

Inorganic mercurial fungicides are probably the most toxic of the fungicides. Mercury's fungicidal properties and toxicity to animals are due in part to the degree of association of divalent mercury ions, which are toxic to all forms of life. As a result, no mercury residues are permitted in foods or feed.

7.11 The most generally toxic of the fungicides are the
     inorganic _____ compounds.

- -

     mercury

There were a host of organic mercury compounds developed over the past 30 years, which have been replaced by other organic fungicides. They bore names such as ceresan and PMA, and were used for seed treatments, turf diseases and dormant sprays on fruit trees.

All mercurial fungicides have lost their registrations and have been discontinued by their manufacturers. Except for a few very specialized applications they are no longer used in the U. S. The basis for these decisions focuses on their toxicity to warm-blooded animals and accumulation of mercury in the environment.

7.12 The mercurial fungicides are no longer used because of
     their _____ to warm blooded animals and its
     _____ in the environment.

- -

     toxicity   accumulation

ORGANIC FUNGICIDES

Over the past 40 years many synthetic sulfur and other organic fungicides have been developed to replace the harsh, less selective inorganic materials. Most of them have had no measurable build-up effect on the environment after many years of use. The first of the organic sulfur fungicides was discovered in 1931; this fungicide, thiram, was followed by many others. Then came other new classes, the dithiocarbamates and dicarboximides (zineb and captan) introduced in 1943 and 1949, respectively. Since then, the organic synthesists have literally left the barn doors open, with now more than 140 fungicides of all classes in use and in various stages of development.

The outstanding qualities of the newer organic fungicides are their extreme efficiency, that is, smaller

quantities are required than of those used in the past and
they usually last longer, and their greater safety to the
crop, to animals, and to the environment.
   Most of the newer fungicides have very low phytotoxicity, many showing at least 10-fold safety factor over
the copper materials. And, most of them are degraded by
soil microorganisms which prevents their accumulation in
soils.

7.13 The newer organic fungicides are distinguished by
     their low _____ and ease in breakdown
     by soil _____.

- -

     phytotoxicity   microorganisms

7-D   DITHIOCARBAMATES

   The dithiocarbamates are the "old reliables." In this
group are thiram, maneb, ferbam, ziram, nabam and zineb,
all developed in the early 1930's and 1940's. Such fungicides probably have greater popularity and use than all the
other fungicides combined, including home gardens. Except
for systemic action, they are employed collectively in every
use known for fungicides. The dithiocarbamates probably act
by being metabolized to the isothiocyanate radical (-N=C=S),
which inactivates the sulfhydryl groups in amino acids and
enzymes within the individual pathogen cells.

7.14 The dithiocarbamates probably are converted to
     _____ which inactivate -SH groups
     in pathogen cells.

- -

     isothiocyanates

THIRAM

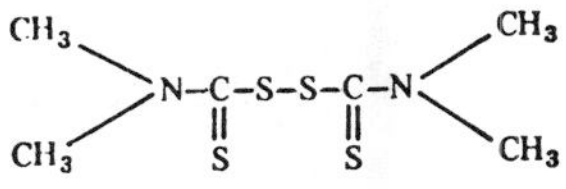

$$CH_3{>}N-C-S-S-C-N{<}CH_3$$

bis(dimethylthio-carbamoyl)
disulfide

MANEB

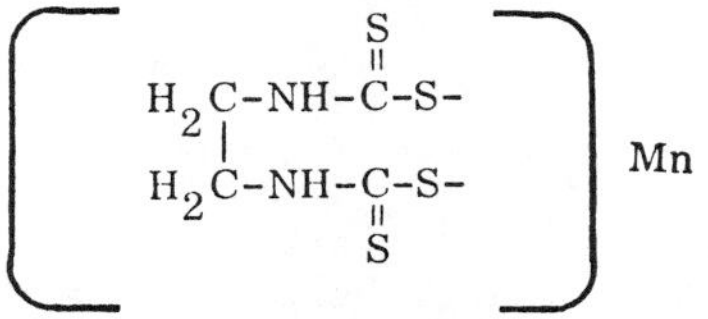

$$H_2C-NH-C-S-$$
$$H_2C-NH-C-S-$$ Mn

manganese ethylenebisdithiocarbamate

NABAM

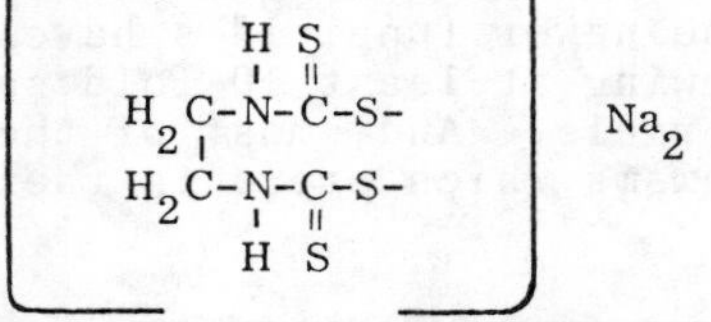

$$H_2C-N-C-S- \\ H_2C-N-C-S-$$

Na_2

disodium ethylenebisdithiocarbamate

FERBAM

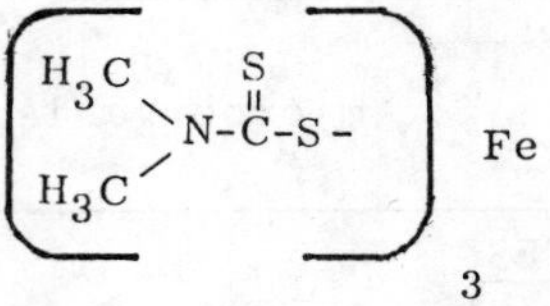

$$H_3C \\ \quad\ \ N-C-S- \\ H_3C$$

Fe

3

ferric dimethyldithiocarbamate

ZIRAM

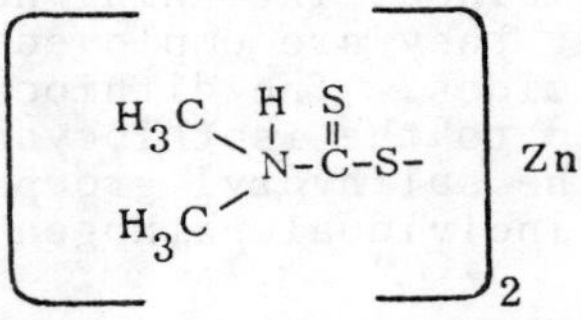

$$H_3C \\ \quad\ \ N-C-S- \\ H_3C$$

Zn

2

zinc dimethyldithiocarbamate

ZINEB

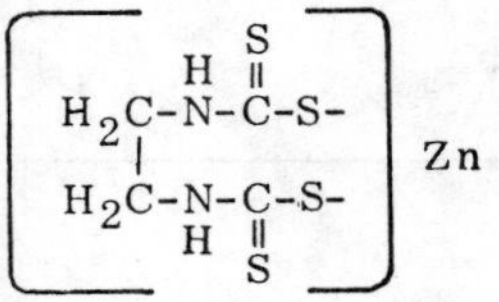

$$H_2C-N-C-S- \\ H_2C-N-C-S-$$

Zn

zinc ethylenebisdithiocarbamate

Chelation: Some of the metals required by the higher plants and fungi in trace amounts assist enzymes in conducting their routine duties of metabolism. The metal may be active in this role as a *chelate* with the biological component. A chelate is an organic ring structure composed of the metal atom linked into the ring by nitrogen, oxygen or sulfur.

7.15 Enzymes in plants and fungi utilize metals in the
     form of _____ to aid enzymes in
     metabolism.

- -

     chelates

     Chelates then become powerful and essential entities
in the metabolic processes of plants, particularly as
they involve enzymes. One of the generally accepted
theories to explain the fungicidal activity of copper, mercury, cadmium and other heavy metals, is the formation of
chelates within the fungal cells, which, in turn, disrupt
protein synthesis and metabolism. And since the most critical proteins of cells are enzymes, DNA, and RNA, it can be
imagined that metals required in trace amounts appearing in
abundance or excessive quantities would be equivalent to the
introduction of potent poisons in the cells.
     If we accept the chelation theory as explaining the
mode of action of the heavy metal fungicides, both organic
and inorganic, and the formation of isothiocyanates from
the dithiocarbamate molecules, the potency of the heavy
metal dithiocarbamates becomes evident.
     Also, there are certain fungicides which are in them-
selves chelating agents. They attach to the scarce metal
components, such as Fe, Mg and Zn, within the cells literally robbing them of essential materials.
     In summary, chelation of heavy metals plays an impor-
tant role in both the life and death of cells.

7.16 Chelates are organic ring structures having
     _____ atoms linked into the ring by
     _____, _____, or _____.

- -

     metal   N, O, or S

7-E THIAZOLES

     Ethazol belongs to the thiazoles, a class of compounds
that offers a surprising chemical disposition. The 5-
membered ring of the thiazoles is cleaved rather quickly
under soil conditions to form either the fungicidal isothiocyanate (-N=C=S) or a dithiocarbamate, depending on the
structure of the parent molecule. Ethazol is used only as
a soil fungicide, and as such, is exposed to the ring-
cleavage just mentioned. To determine the probable mode of
action, review the action of dithiocarbamates.

7.17 Ethazol is probably cleaved to form a dithiocarbamate
     which in turn is converted to an _____
     which inactivates _____ groups in enzymes.

----

     isothiocyanate   -SH

ETHAZOL (Terrazole[R])

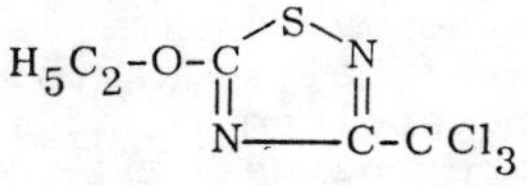

$$H_5C_2-O-C \overset{S}{\underset{N}{\diagdown}} \underset{C-C\,Cl_3}{\overset{N}{\|}}$$

5-ethoxy-3-trichloromethyl-1,2,4-thiadiazole

7-F TRIAZINES

The triazine moiety, which we see frequently in herbicides, is found only once in the fungicides. Anilazine
was introduced in 1955, and has received wide use for control of potato and tomato leafspots and turfgrass diseases
around the home.

ANILAZINE (Dyrene[R])

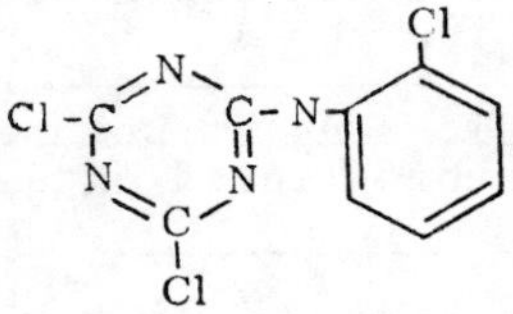

2,4-dichloro-6-(o-chloroanilino)-s-triazine

7-G SUBSTITUTED AROMATICS

The substituted aromatics belong in a somewhat arbi-
trary classification assigned to the simple benzene derivatives that possess long-recognized fungicidal properties.
     Hexachlorobenzene was introduced in 1945, and is used
as a seed treatment and as a soil treatment to control
stinking smut of wheat. Pentachlorophenol (PCP) has been
used since 1936 as a wood preservative and seed treatment.
Pentachloronitrobenzene (PCNB) was introduced in the 1930s
as a fungicide for seed treatment and selected foliage applications. It is also used as a soil treatment to control
the pathogens of certain damping-off diseases of seedlings.

Chlorothalonil is a very useful, broad-spectrum foliage protectant fungicide made available in 1964. And chloroneb, developed in 1965, is heavily used for cotton seedling and turf diseases.

HEXACHLOROBENZENE

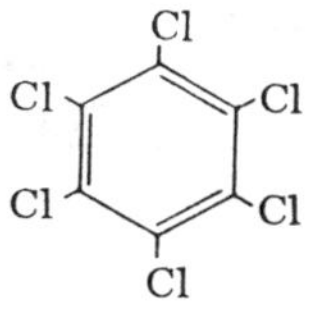

1,2,3,4,5,6-hexachlorobenzene

PCP

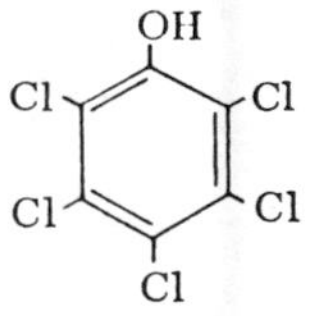

pentachlorophenol

PCNB (Terraclor[R])

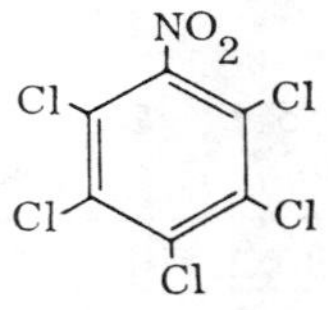

pentachloronitrobenzene

CHLOROTHALONIL (Bravo[R])

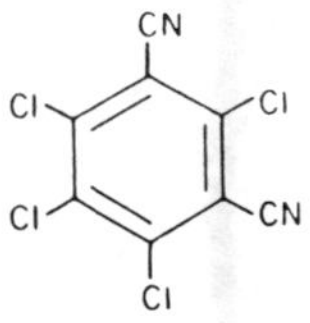

tetrachloroisophthalonitrile

CHLORONEB (Demosan[R], Tersan[R])

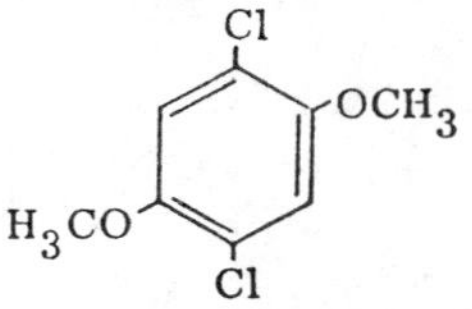

1,4-dichloro-2,5-dimethoxybenzene

Dicloran (DCNA) is a highly useful fungicide against *Botrytis*, *Monilinia*, *Rhizopus*, *Sclerotinia* and *Sclerotium* species, on a wide range of fruits and vegetables.

DICLORAN (DCNA, Botran[R])

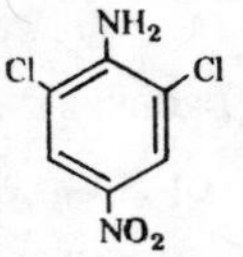

2,6-dichloro-4-nitroaniline

The substituted aromatics are diverse in their modes of action. Being generally fungistatic, they reduce growth rates and sporulation of fungi, probably by combining with $-NH_2$ or $-SH$ groups of metabolically essential compounds.

7.18 The substituted aromatic fungicides are probably
     effective through their attachment to essential bio-
     chemicals containing _____ or _____ groups.

- -

     $-NH_2$ or $-SH$

7-H <u>DICARBOXIMIDES</u> (SULFENIMIDES)

Three extremely useful foliage protectant fungicides belong to this group: captan, which appeared in 1949; folpet, 1962; and captafol (Difolatan[R]), 1961. They were used primarily as foliage dusts and sprays on fruits, vegetables and ornamentals. All registered uses of captan on food crops have been cancelled by the EPA.

CAPTAN

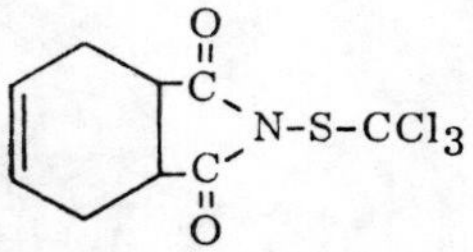

N-(trichloromethylthio)-4-cyclohexene-1,2-dicarboximide

FOLPET (Phaltan[R])

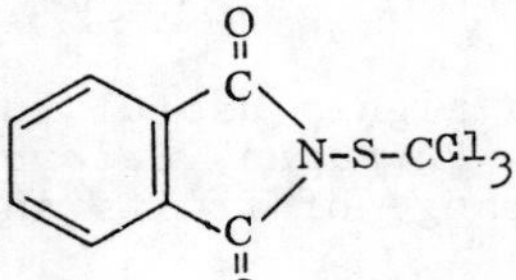

<u>N</u>-trichloromethylthiophthalimide

CAPTAFOL (Difolatan®)

$$\begin{array}{c}
\text{O} \\
\| \\
\text{C} \\
\diagdown \\
\quad\quad\text{N-S-C-CH} \\
\diagup \\
\text{C} \\
\| \\
\text{O}
\end{array}
\quad
\begin{array}{c}
\text{Cl Cl} \\
| \ | \\
\text{C-CH} \\
| \ | \\
\text{Cl Cl}
\end{array}$$

tetrachloroethylmercaptocyclohexenedicarboximide

The dicarboximides were probably the most used fungi-cides for lawn and garden, effective as seed treatments and as protectants against mildews, late blight, and many other diseases.

Many compounds containing the -SCCl$_3$ moiety are fungi-toxic, indicating that group as a *toxophore*. Fungitoxicity of the dicarboximides is apparently nonspecific and is not a result of a single mode of action. Their lethal action is probably due to the inhibition of synthesis of amino compounds and enzymes containing the sulfhydryl (-SH) radical.

7.19 As with the substituted aromatics, the dicarboximides inhibit the production of essential compounds con-taining _____ and _____ groups.

- -

-NH$_2$ -SH

SYSTEMIC FUNGICIDES

Systemic fungicides are those absorbed by the plant and carried by translocation through the cuticle and across leaves to the growing points. It has only been in recent years that successful systemics have been marketed, and only a few are yet available. Most systemic fungicides have eradicant properties which stop the progress of exist-ing infections. They are *therapeutic* in that they cure plant diseases. A few of the systemics can be applied as soil treatments and are slowly absorbed through the roots to give prolonged disease control.

These systemics offer much better control of diseases than is possible with a protectant fungicide that requires uniform application and remains essentially where it is sprayed onto the plant surfaces. There is, however, some redistribution of protective fungicidal residues on the surfaces of sprayed or dusted plants, giving them longer residual activity than would be expected.

7.20 Systemic fungicides are translocated through
_____ and across _____.
- -

    cuticle   leaves

7.21 Systemics offer better disease control than the
_____ which are effective only where
they are applied to the plant.
- -

    protectants

    Those systemics currently in commercial use will be
mentioned in their chronological order of appearance.

7-I OXATHIINS

    The two systemic fungicides in this group, carboxin
and oxycarboxin, both introduced in 1966, were the first
of the systemics to succeed in practice. They are used as
seed treatments for the cereal crops, particularly those
affected by embryo-infecting smut fungi, and they have
potential for other uses. They are selectively toxic to
the smuts, rusts, and to *Rhizoctonia (Thanatephorus)*,
organisms belonging to the Basidiomycetes. The apparent
mode of action of the oxathiins begins with their selective
concentration in the fungal cells, followed by the inhibition of succinic dehydrogenase, an important enzyme to
respiration in the mitochondrial systems.

7.22 The oxathiins are selective against _____ and
_____ fungi, by inhibition of _____
_____ in the mitochondria.
- -

    smut and rust   succinic dehydrogenase

CARBOXIN (Vitavax^R)

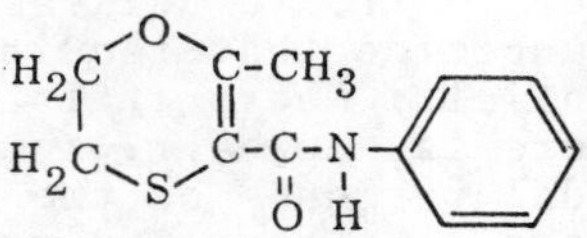

2,3-dihydro-5-carboxanilido-
6-methyl-1,4-oxathiin

OXYCARBOXIN (Plantvax^R)

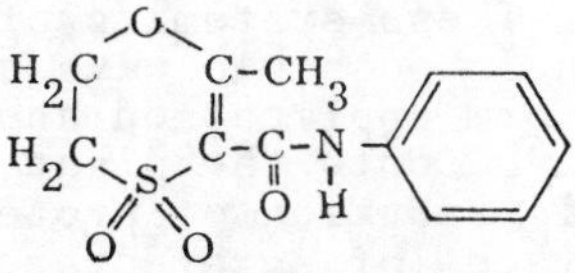

2,3-dihydro-5-carboxanilido-
6-methyl-1,4-oxathiin-4,4-dioxide

7-J BENZIMIDAZOLES

The benzimidazoles, represented by benomyl and thiabendazole (TBZ), were introduced in 1968, and have received
wide acceptance as systemic fungicides against a broad
spectrum of diseases. Benomyl has the widest spectrum of
fungitoxic activity of all the newer systemics, including
the *Sclerotinia* spp., *Botrytis* spp., *Rhizoctonia* spp.,
powdery mildews, and apple scab. Thiabendazole (TBZ) has
a similar spectrum of activity to that of benomyl. Thiophanate, (Topsin[R] E), whose structure is not shown, is not
a benzimidazole. But because it is converted to this
group's structure when metabolized, it belongs to them.
Thiophanate has a fungitoxic activity similar to that of
benomyl. All three compounds have been used in foliar applications, seed treatment, dipping fruit or roots, and soil
applications. Their mode of action appears to be the induction of abnormalities in spore germination, cellular multiplication and growth as a result of their interference in
mitosis and in the synthesis of nucleic material (DNA).

7.23 The benzimidazole fungicides act by interfering in
_____ and in the synthesis of the nucleic
material _____.

- -

mitosis   DNA

BENOMYL (Benlate[R])   THIABENDAZOLE (TBZ)

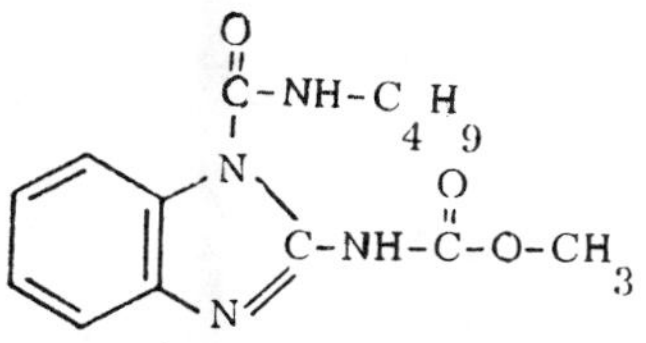

methyl-1-(butylcarbamoyl-2-
benzimidazolecarbamate

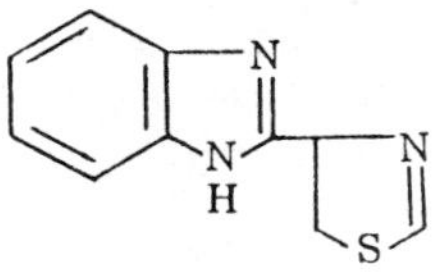

2-(4'-thiazoyl) benzimidazole

Imazalil is related to benomyl and thiabendazole in
that it is an imidazole compound without the benzene ring.
Imazalil is also systemic used as both a curative and as a
preventative. It is currently registered in the U. S. for
cotton, wheat, and barley seed treatment, and for post-
harvest use on citrus.

IMAZALIL (Bromazil[R], Deccozil[R])

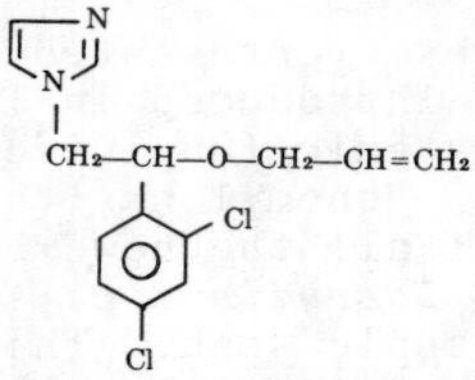

1-(2-(2,4-dichlorophenyl)-2-
(2-propenyloxy) ethyl)-1H-imidazole

7-K PYRIMIDINES

The pyrimidine systemic fungicides appeared in the
late 1960s, and include dimethirimol, ethirimol and bupirimate. They are very active against specific types of powdery mildews; for instance, dimethirimol works well on
cucurbits; ethirimol is for cereals and other field crops,
and bupirimate controls powdery mildews on apples and greenhouse roses.

DIMETHIRIMOL (Milcurb[R])

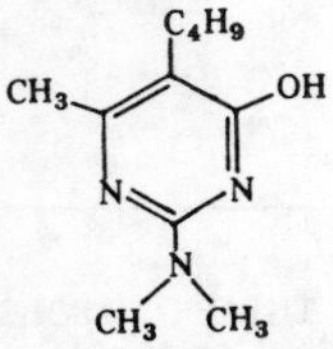

5-n-butyl-2-dimethylamino-4-
hydroxy-6-methylpyrimidine

BUPIRIMATE (Nimrod[R])

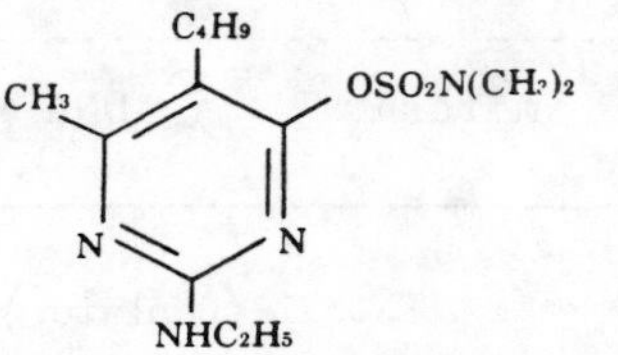

5-butyl-2-ethylamino-6-methyl-
pyrimidin-4-yl-dimethylsulfamate

ETHIRIMOL (Milcurb[R] Super)

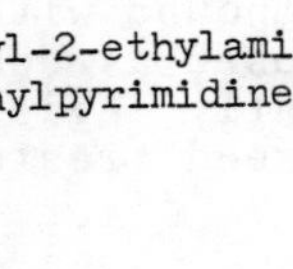

5-butyl-2-ethylamino-4-hydroxy-
6-methylpyrimidine

7-L   ORGANOPHOSPHATES

IBP is the first organophosphate developed as a fungicide. It is not only a protectant but also a systemic,
eradicating fungicide with some insecticidal properties.
It is registered only for rice outside the U. S. A more
recent organophosphate is fosetyl-aluminum, usually referred to as fosetyl. It, too, is a systemic foliar fungicide,
registered only on ornamentals. Experimentally it is being
tested on a broad variety of fruits and vegetables. The
mode of action for the organophosphates is uncertain, but
tentatively they are considered to interfere in phospholipid
synthesis of the invading microorganism.

IBP (Kitazin^R)   FOSETYL (Aliette^R)

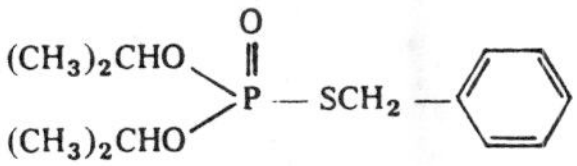

O,O-diisopropyl-S-benzyl   aluminum tris-O-ethyl phosphonate
thiophosphate

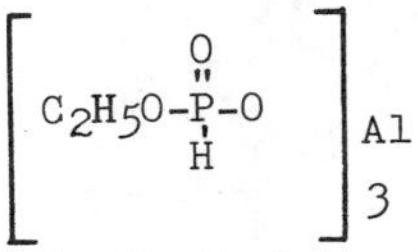

7-M   ACYLALANINES

Another new group of systemic fungicides are the
acylalanines, which include metalaxyl and furalaxyl. They
are effective against soil-borne diseases caused by *Pythium*
and *Phytophthora* and foliar diseases caused by the Oomycetes
(downy mildews), and also show promise as foliar, soil, and
seed treatments for agricultural crops. They offer systemic
and curative as well as residual-protectant activity, but
are not effective against the Ascomycetes, Basidiomycetes
and Fungi Imperfecti.

METALAXYL (Ridomil^R, Subdue^R)   FURALAXYL (Fongarid^R)

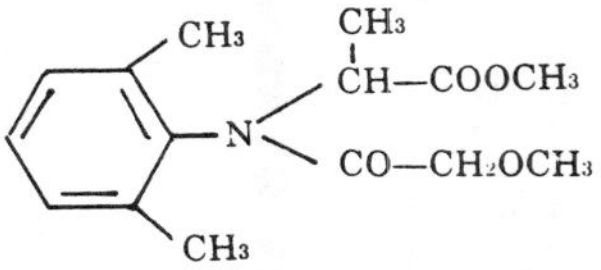

N-(2,6-dimethylphenyl)-N-(methoxy=   methyl N-2,6-dimethylphenyl-N-
acetyl)-alanine methyl ester   furoyl-(2)-alaninate

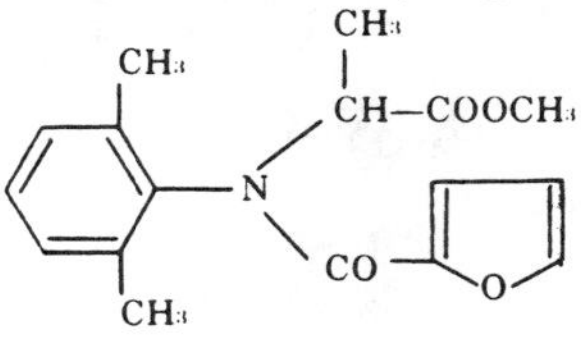

7-N TRIAZOLES

The triazoles are a new class of systemic fungicides, that have both protective and curative actions. They are broad spectrum in their action controlling powdery mildew and rusts among others. Of the four compounds being marketed, triadimefon is the only one registered on food crops. The other three, propiconazole, triadimenol and bitertanol are under experimental use permits for food and feed crops. The triazoles are among the newest and most promising fungicides of the future, and are believed to inhibit the synthesis of sterols.

TRIADIMEFON (Bayleton[R])

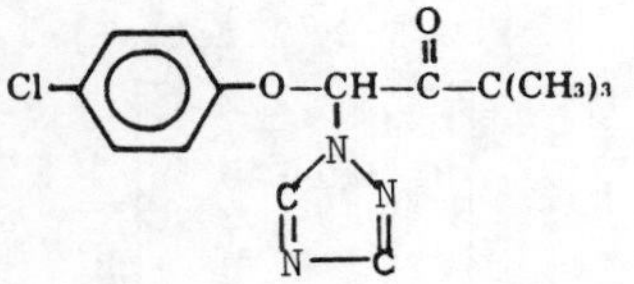

1-(4-chlorophenoxy)-3,3-dimethyl-
1-(1HO1,2,4-triazol-1-yl)-2-butanone

PROPICONAZOLE (Tilt[R])

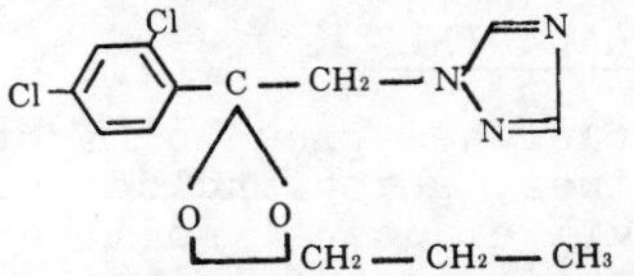

1-(2-(,4-dichlorophenyl)4-propyl-1,3-
dioxolan-2-ylmethyl)-1H-1,2,4-triazole

TRIADIMENOL (Baytan[R])

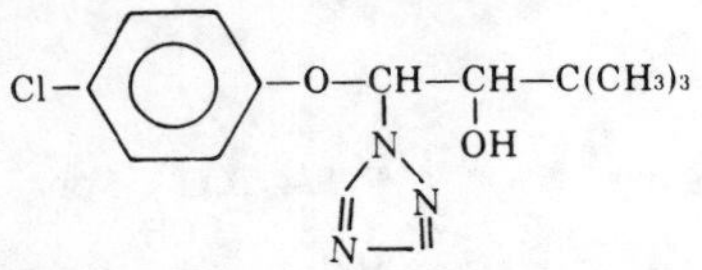

beta-(4-chlorophenoxy)-alpha-(1,1-
dimethylethyl) 1H-1,2,4-triazole-1-ethanol

BITERTANOL (Baycor[R])

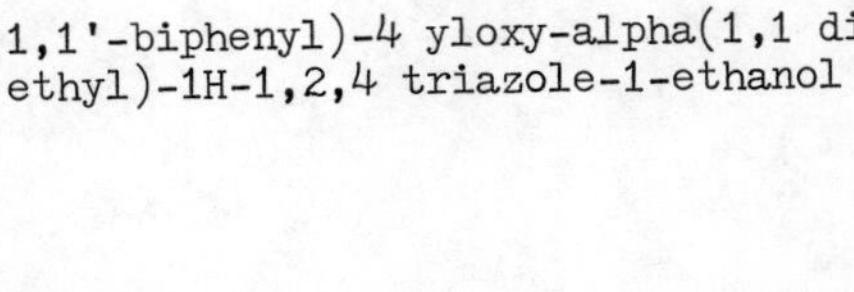

beta-((1,1'-biphenyl)-4 yloxy-alpha(1,1 dimethylethyl)-1H-1,2,4 triazole-1-ethanol

7-O PIPERAZINES

The piperazine systemic fungicides include only one candidate at this time, triforine, which is amazingly effective against powdery mildew on literally any host. Additionally, it is active against scab on fruits, rust and black spot on ornamentals, several diseases of vegetables and many others including powdery mildew. It first appeared in the mid-1970s and is registered for powdery mildew on greenhouse roses, brown rot blossom blight on peaches, and mummy berry disease on blueberries. Triforine is believed to interfere in the synthesis of sterols.

TRIFORINE (Funginex^R)

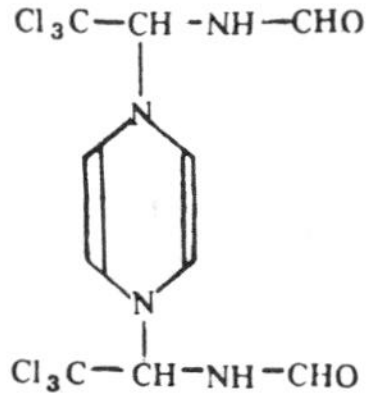

Cl_3C—CH -NH—CHO

Cl_3C— CH—NH—CHO

N,N'-(1,4-piperazinediyl-bis(2,2,2-
trichloroethylidene))-bis-(formamide)

7-P IMIDES

A second group of dicarboximides, the imides, which are structurally significantly different from the originals and systemic, appeared a decade later. For the time, these include procymidone, iprodione, and vinclozolin. They are particularly effective against *Botrytis*, *Monilia* and *Sclerotinia*. Iprodione is also active against *Alternaria*, *Helminthosporium*, *Rhizoctonia*, *Corticium*, *Typhula* and *Fusarium* spp.

IPRODIONE (Rovral^R)

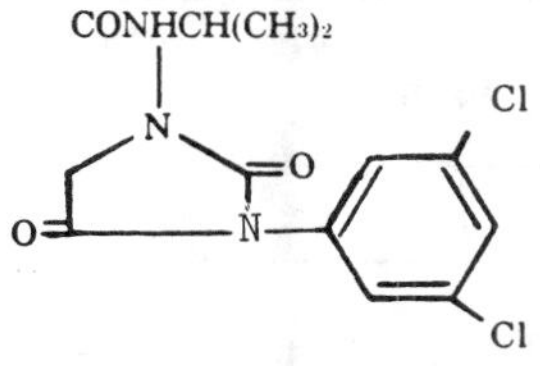

CONHCH(CH₃)₂

VINCLOZOLIN (Ronilan^R)

3-(3,5-dichlorophenyl)-N-(1-methyl-
ethyl)-2,4-dioxo-1-imidazolidine=
carboxamide

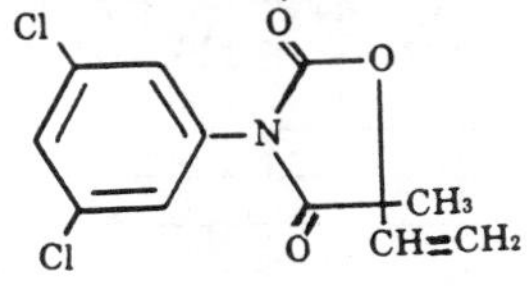

3-(3,5-dichlorophenyl)-5-ethenyl-
5-methyl-2,4-oxazolidinedione

PROCYMIDONE (Sumilex^R)

N-(3,5-dichlorophenyl)-1,2-dimethyl=
cyclopropane-1,2-dicarboximide

The systemic fungicides cover susceptible foliage and
flower parts more efficiently than protectant fungicides
because of their ability to translocate through the cuticle
and across leaves. They bring into play the ideal method
of disease control by attacking the pathogen at its site
of entry or activity, and reducing the risk of contami-
nating the environment by frequent broadcast fungicidal
treatments. Undoubtedly, as these new and more selective
systemic molecules are synthesized, they will gradually
replace the protectants which still comprise the bulk of
our fungicidal arsenal.

Resistance to chemicals other than the heavy metals
occurs commonly in fungal and on rare occasions in bacterial plant disease pathogens. Several growing seasons
after a new fungicide appears, it becomes noticeably less
effective against a particular disease. As our fungicides
become more specific for selected diseases, we can expect
the pathogens to become resistant. This can be attributed
to the singular mode of action of a particular fungicide
which disrupts only one genetically controlled process in
the metabolism of the pathogen. The result is that resistant populations appear suddenly, either by selection of
resistant individuals in a population or by a single-gene
mutation. Generally, the more specific the site and mode of
fungicidal action, the greater the likelihood for a pathogen to develop a tolerance to that chemical. As of 1985,
there are more than 100 species of plant pathogens resistant
to one or more fungicides.

7.24 Why haven't plant disease pathogens developed resis-
     tance to the older heavy metal fungicides?

     _____

     _____

- -

     They have no single mode of action. Rather they would
     be considered "broad-side" fungicides. Thus no single
     genetically-controlled process to be selected, resulting
     in resistance.

7.25 Present two logical reasons why systemic fungicides
     are superior to protectants.
     (1) _____
     (2) _____
----

     (1) more efficient - attack after penetration during
         early stages of infection
     (2) don't contaminate environment as much as
         protectants

7.26 What are the two basic routes for development of
     resistance to a fungicide by a plant pathogen?
     A. _____
     B. _____

     A. Selection of resistant individuals in a population
     B. Single-gene mutation

7-Q PYRIMIDINES

     The pyrimidine fungicides also contain but one of
prominence, fenarimol.   Their probable mode of action is
the inhibition of sterol synthesis in the demethylation at
the C-14 position.   As a group they exert their action
against a broad spectrum of pathogens, except the Oomycetes
(downy mildew).   Rubigan[R] is currently registered only for
turf, but is being tested on deciduous fruits, grapes,
cucurbits, grapes, ornamentals and peanuts.

FENARIMOL (Rubigan[R])

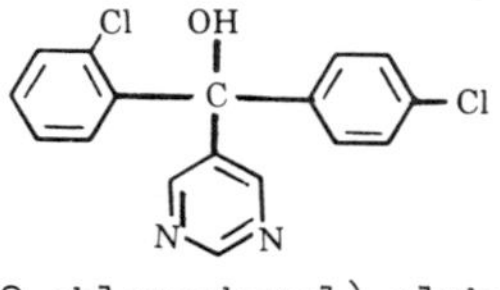

alpha(2-chlorophenyl)-alpha-(4-
chlorophenyl)-5-pyrimidinemethanol

7-R CARBAMATES

     The only member of this group is propamocarb, a sys-
temic soil and foliar fungicide.   It appeared in 1978 and
is being developed in the U. S. as a turf and ornamental
material.   Experimentally it is being tested on vegetables
and tobacco, and is particularly effective against Pythium
blight on turf.

PROPAMOCARB (Banol[R], Prevex[R])

$$\begin{matrix} CH_3 \\ \\ CH_3 \end{matrix} \Big\rangle NC_3H_6NHCOOC_3H_7 \cdot HCl$$

propyl (3-(dimethylamino)propyl)
carbamate monohydrochloride

7-S FUMIGANTS

As with the insecticides, there are several highly
volatile, small molecule   fungicides that have fumigant
action. They are unrelated chemically, but are handled
similarly and will be dealt with as a single class in this
book. Chloropicrin was mentioned before as a warning agent
in grain fumigants, but it is also an ideal fumigant itself;
it controls fungi, insects, nematodes, and weed seeds in the
soil.

Methyl bromide, also listed as a fumigant insecticide,
is equally effective against fungi, nematodes, and weeds.
Methylisothiocyanate (MIT) is closely related to the dithiocarbamates and has a similar mode of action against fungi,
nematodes, and weeds. Metam-sodium was listed, but not
illustrated, in the dithiocarbamate fungicides, where it
belongs chemically. Metam-sodium decomposes in the soil to
yield methylisothiocyanate (MIT).

METHYL BROMIDE

CH_3Br

bromomethane

METAM-SODIUM (SMDC, Vapam[R])

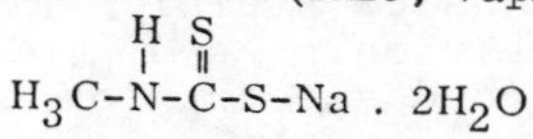

$$\begin{matrix} H & S \\ | & || \\ H_3C-N-C-S-Na \end{matrix} \cdot 2H_2O$$

sodium $\underline{N}$-methyldithiocarbamate dihydrate

CHLOROPICRIN

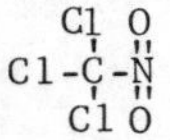

$$\begin{matrix} Cl & O \\ | & || \\ Cl-C-N \\ | & || \\ Cl & O \end{matrix}$$

trichloronitromethane

MIT

$CH_3-N=C=S$

methylisothiocyanate

7.27 MIT and Metam-sodium have the same mode of action as
     the dithiocarbamates, the inactivation of _____
     groups in amino acids, proteins, and enzymes.

- -

     -SH

7-T ANTIBIOTICS

The antibiotic fungicides are substances produced by
microorganisms, which in very dilute concentrations inhibit
growth and even destroy other microorganisms. To date

several hundred antibiotics have been reported to have
fungicidal activity, and the chemical structures of about
half of these are already known. (See also the section on
Antibiotics in the Insecticides chapter.)

7.28 Antibiotic fungicides which destroy and inhibit growth
     of fungi, are produced by _____.

- -

     microorganisms

     The largest source of antifungal antibiotics is the
actinomycetes, a group of the lower plants. Within the
actinomycetes is one amazing species, *Streptomyces griseus*,
from which we obtain streptomycin and cycloheximide.
     Streptomycin is used as dusts, sprays and seed treat-
ment for the control mostly of bacterial diseases such as
blight on apples and pears, soft rot on leafy vegetables
and some seedling diseases. It is also effective against
a few certain fungal diseases.
     The mode of action of streptomycin is not clearly
understood, but it probably interferes in the synthesis
of proteins, specifically RNA and DNA, vital nucleic acids
of living cells. Despite the evidence of antibiotic-
resistant strains, streptomycin has a place in the control
of some bacterial diseases, and the tetracyclines may well
play an important part in controlling some mycoplasma-like
diseases of plants. Because of the scientific principles
behind its development, resistance to antibiotics by both
bacteria and mycoplasmas is inevitable.

STREPTOMYCIN

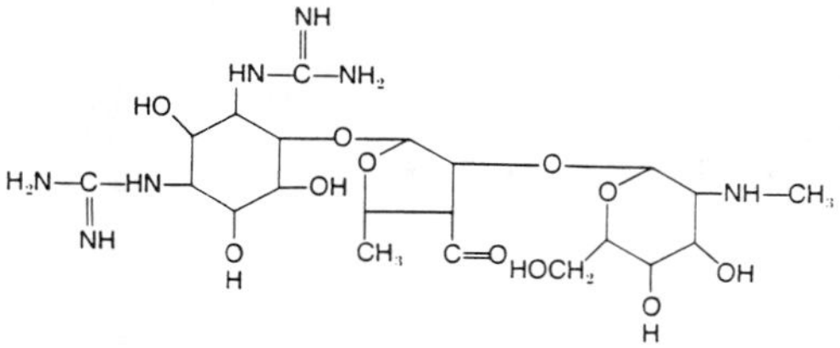

2,4-diguanidino-3,5,6-trihydroxycyclohexyl-5-deoxy-2-0-(2-
deoxy-2-methylamino-a-glucopyranoxyl)-3-formyl pentofuranoside

7.29 Streptomycin probably acts by inhibiting the synthesis
     of the vital nucleic acids _____ and _____.

     RNA and DNA

    Cycloheximide is a smaller, less complicated anti-
biotic, about which more is understood. First, cycloheximide is toxic to a wide range of organisms, including
yeasts, filament-forming fungi, algae, protozoa, and higher
plants and especially mammals (See Panel M). Surprisingly
it is inactive against bacteria. As a matter of specula-
tion, this may result from failure of bacteria to absorb
it. Cycloheximide causes growth inhibition of yeasts and
filament-forming fungi by inhibiting protein and DNA synthesis. Then, with a fair amount of confidence, we can say
that both streptomycin and cycloheximide act by inhibiting
the synthesis of nucleic acids. Cycloheximide is not only
toxic to fungi, but it can also kill plants and has an oral
LD_{50} in the rate of 2.5 mg/kg, by far the most toxic of
fungicides.

7.30 Streptomycin and cycloheximide both inhibit growth of
     microorganisms by inhibiting the synthesis of
     _____.

----

     nucleic acids

    Cycloheximide was introduced as a fungicide in 1948,
and has since become popular in the control of powdery mildew, rusts, turf diseases, and certain blights. It is best
known commercially under the name Acti-dione[R]. It cannot be
purchased for home and garden use because of its high acute
toxicity.

CYCLOHEXIMIDE (Acti-dione[R])

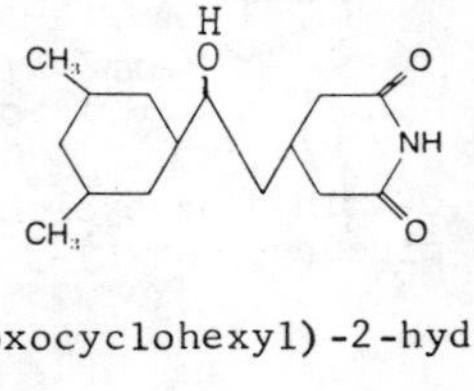

β[2-(3,5-dimethyl-2-oxocyclohexyl)-2-hydroxyethyl]-glutarimide

Blasticidin-S, discovered in 1955, is produced by the fermentation of *Streptomyces griseochromogenes*. Kasuga-mycin[R], introduced in 1963, is formed by *Streptomyces kasugaenis*, and Polyoxin-B[R] is extracted from *Streptomyces oacaoi*. All are Japanese contributions to the systemic antibiotics. The first two are effective against rice blast, and the latter against rice sheath blight. None of these are illustrated. Their mode of action is the same as cycloheximide.

7-U DINITROPHENOLS

The dinitrophenols should be familiar by now, because of their previous appearances as insecticide-ovicides and herbicides. Their mode of action continues the same here, uncoupling oxidative phosphorylation in cells with an attendant upset of the energy systems within the cells. The statement holds true in all cases.

7.31 The dinitrophenols act by uncoupling_____
_____ in most living cells.

- -

oxidative phosphorylation

One dinitrophenol has been used since the late 1930s, dinocap, both as an acaricide and for powdery mildew on a number of fruit and vegetable crops. Dinocap undoubtedly acts in the vapor phase since it is quite effective against powdery mildews whose spores germinate in the absence of water.

DINOCAP (Karathane[R])

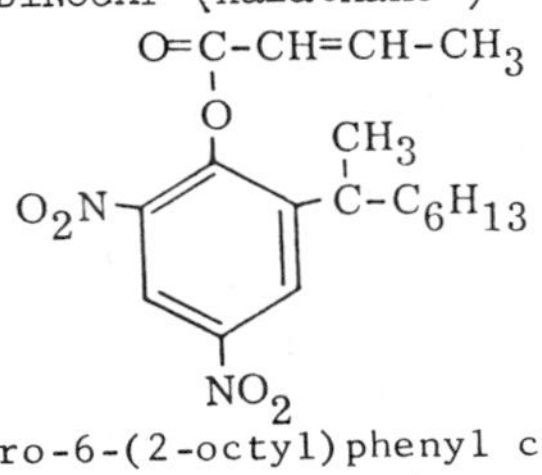

O=C-CH=CH-CH$_3$
|
O   CH$_3$
| |
O$_2$N   C-C$_6$H$_{13}$

NO$_2$

2,4-dinitro-6-(2-octyl)phenyl crotonate

7-V QUINONES

The quinones are a fascinating chemical group, because there are practically unending numbers of molecules that are potential fungicides. Chloranil, the first of these to appear (1937), was used heavily as a seed treatment and for foliar application until the dicarboximides appeared (captan, etc.). The most popular of this group, however, was dichlone. It is used on fruit and vegetable crops and in ponds, to control blue-green algae. Dichlone affects cellular respiration in fungi by attaching to the -SH groups in enzymes, and indirectly uncoupling oxidative phosphorylation.

7.32 The quinones attach to -SH groups in enzymes, ulti-
     mately preventing _____ _____.

- -

oxidative phosphorylation

DICHLONE (Quintar^R)   CHLORANIL (Spergon^R)

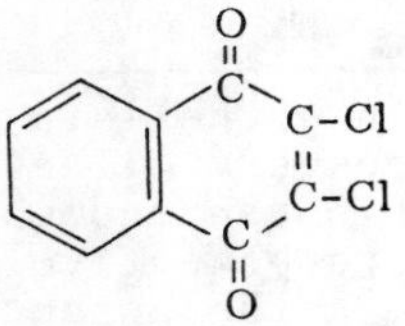

2,3-dichloro-1,4-naphthoquinone   2,3,5,6-tetrachloro-1,4-benzoquinone

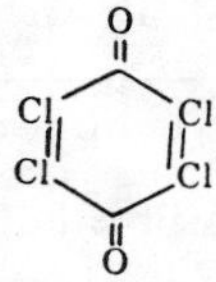

7-W ORGANOTINS

The organotins were first introduced in the mid-1960s, ten years after their fungicidal properties had been dis-covered. In general, the trialkyl derivatives are highly fungicidal, but also phytotoxic. The triaryl (3-ring) com-pounds are suitable for protective use, and also have acar-icidal properties. Among these are fentin hydroxide (Du-Ter^R), triphenyltin acetate (Brestan^R), and fentin chlor-ide (Tinmate^R) which is also formulated for snail control in fishponds (Aquatin^R). Only the first is illustrated. The tri-substituted tins block oxidative phosphorylation, as mentioned in the insecticides.

7.33 In the organotin compounds, which group is not
     phytotoxic, the trialkyl or triaryl? _____

- -

   triaryl

FENTIN HYDROXIDE (Du-Ter^R)

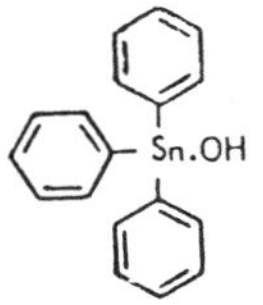

triphenyltinhydroxide

7-X ALIPHATIC NITROGENS

   Dodine, introduced in the middle 1950s, is effective
in controlling certain diseases such as apple and pear scab
and cherry leaf spot. It has disease-specificity, and
slight systemic qualities. Its mode of action is not
totally clear, but it is taken up rapidly by fungal cells,
causing leakage in these cells possibly by alteration in
membrane permeability. The guanadine nucleus of dodine is
also known to inhibit the synthesis of RNA.

7.34 Dodine causes affected cells to _____,
     and is known to interfere in the synthesis of
     _____.

- -

   leak   RNA

DODINE (Cyprex^R, Melprex^R)

$$H_{25}C_{12}-NH-\underset{\underset{NH}{\|}}{C}-NH_2-CH_3COOH$$

n-dodecylguanidine acetate

7.35 How many of these structures representing major
     fungicide classes can you identify?

     Inorganic copper____; dithiocarbamate____; thiazole____;
     carbamate____; quinone____; antibiotic____; oxathiin
     ____; benzimidazole____; dicarboximide____; substituted
     aromatic____.

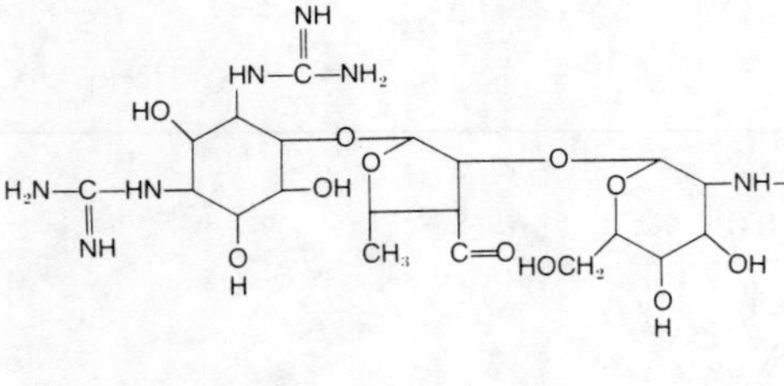

1 — antibiotic structure (streptomycin-type) with guanidine groups:

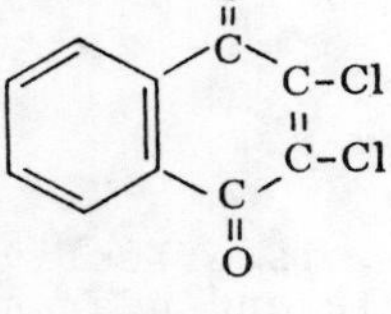

NH
HN—C—NH$_2$
HO
H$_2$N—C—HN
NH
H
O
OH
OH
CH$_3$
C=O HOCH$_2$
OH
OH
H
NH—CH$_3$

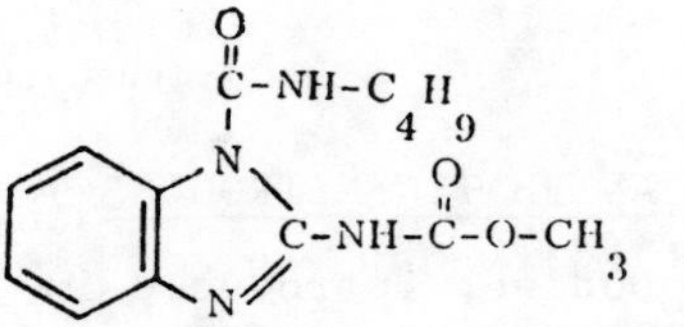

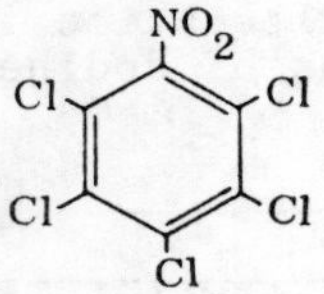

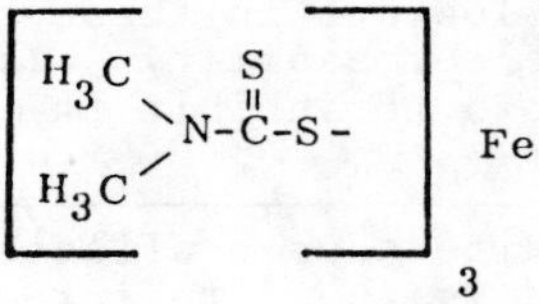

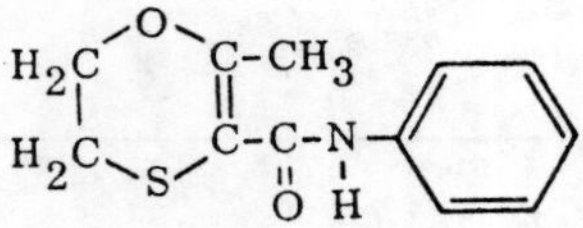

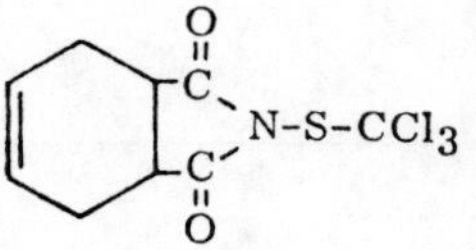

1

$$CuSO_4(N_2H_5)_2SO_4$$

2

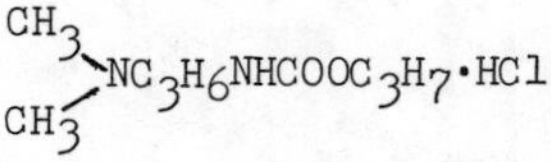

$$\begin{array}{c}CH_3\\ \diagdown\\ CH_3\end{array} NC_3H_6NHCOOC_3H_7 \cdot HCl$$

3

4 — phthaloyl dichloride type:

O
‖
C—Cl
‖
C—Cl
‖
O

4

5 — benzimidazole:

O
‖
N—C—NH—C$_4$H$_9$
O
‖
C—NH—C—O—CH$_3$
N

5

6 — substituted aromatic:

NO$_2$
Cl   Cl
Cl   Cl
Cl

6

7 — dithiocarbamate:

$$\left[\begin{array}{c} H_3C \\ \diagdown \\ H_3C \end{array} N\text{--}\overset{\displaystyle S}{\overset{\|}{C}}\text{--}S\text{--} \right]_3 Fe$$

7

8 — oxathiin:

H$_2$C   O   C—CH$_3$
H$_2$C   S   C—C—N
            ‖ H
            O
(phenyl)

8

9 — dicarboximide:

O
‖
C
N—S—CCl$_3$
C
‖
O

9

10 — thiazole:

H$_5$C$_2$—O—C   S   N
            N   C—CCl$_3$

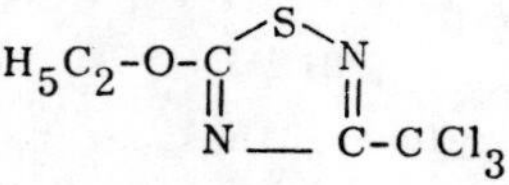

10

----

inorganic copper __2__ ; dithiocarbamate __7__ ;
thiazole __10__ ;   carbamate __3__ ; quinone __4__ ;
antibiotic __1__ ; oxathiin __8__ ; benzimidazole __5__ ;
dicarboximide __9__ ; substituted aromatic __6__ .

UNIT 8
NEMATICIDES

Nematodes are microscopic roundworms that live in soil or water. Many are free-living, whereas others are parasitic on plants or animals. Some species of nematodes inadvertently introduce pathogenic root-invading microorganisms into the plants while feeding. Nematodes may also predispose crop-plant varieties to other disease-causing agents, such as wilts and root rots. In other instances the nematodes themselves cause the disease, disrupting the flow of water and nutrients in the xylem system, resulting in root-knot or deprivation of the aboveground parts, and ultimately causing stunting.

Nematodes are covered with an impermeable cuticle, which provides them with considerable protection. Chemicals with outstanding penetration characteristics are required, therefore, for their control.

The nematicides fall into four groups: (1) halogenated hydrocarbons (some of which were described earlier), (2) organophosphate insecticides, (3) isothiocyanates, and (4) carbamate or oxime insecticides.

8.1   How do nematodes cause disease in plants?

_____

_____

- - - - - - - - - - - - - - - - - - - -

They introduce root-invading microbes; weaken the plants making them susceptible to wilts and root rots; and cause the disease themselves by clogging transport vessels.

8.2   Why are highly-penetrating chemicals required as nematicides?

_____

- -

Cuticle covering of nematodes is impermeable.

The nematicides used most abundantly over the 40-year span from 1945 to 1984 were the volatile halogenated soil fumigants.

To be successful they had to have high vapor pressures to
spread through the soil and contact nematodes in the water
films surrounding soil particles.

8-A HALOGENATED HYDROCARBONS

Discovery of the nematicidal properties of DD and ethylene dibromide (EDB) in 1943 and 1945 effectively launched
the use of volatile nematicides on a field-scale basis.
Previously only seed beds, greenhouse benches, and potting
soil had been treated with the costly soil sterilants chloropicrin, carbon disulfide, and formaldehyde. Not only were
they expensive, but in some instances explosive, and usually
required a surface seal.
DD and EDB were both injected into the soil several days
before planting to kill nematodes, eggs and soil insects.
They differed considerably in uses, soil retention and human
hazard. Now, however, neither compound is available. The
uses of EDB as a soil fumigant have been cancelled, and U.S.
production of DD has stopped, making its use unlikely.

DICHLOROPROPENE-DICHLOROPROPANE (D-D)

$$\underset{\text{HC= CH -CH}_2}{\overset{Cl \qquad Cl}{|\qquad\;\;|}} \quad \text{and} \quad \underset{\text{H}_2\text{C-CH-CH}_3}{\overset{Cl\;Cl}{|\;|}}$$

1,3-dichloropropene and 1,2-dichloropropane

EDB

$Br-CH_2-CH_2-Br$

ethylenedibromide

DBCP was one of the most useful and easily applied
nematicides in agricultural history, becoming indispensable
in grape and citrus production. In 1977 it was announced by
its producers that DBCP had caused sterility in male workers
involved in its manufacture. It was removed from the market
and all registrations eventually cancelled, the last being
on Hawaiian pineapple, cancelled in 1985. Because no satisfactory replacement has been found, nematologists are
still conducting an extensive search for a substitute.

DBCP

$$\underset{\text{CH}_2\text{-CH-CH}_2}{\overset{Br\quad Br\quad Cl}{|\quad\;\;|\quad\;\;|}}$$

dibromochloropropane

Methyl bromide is a first-class, all-around fumigant, lethal to all plant and animal life, hence its classification as a sterilant. The gas is a soil fumigant for control of weeds, weed seeds, nematodes, insects and soil-borne diseases. It is also used for dry-wood termite control above ground, agricultural commodity fumigation, and rodent control. As a nematicide, because of its phytotoxicity, it must be used as a preplant application followed by adequate aeration time. A two-week waiting period after fumigation is an acceptable rule-of-thumb. Methyl bromide is the last remaining halogenated soil fumigant, and its future is uncertain at best.

METHYL BROMIDE

CH_3Br

bromomethane

The mode of action for most of these halogenated nematicides is that of a narcotic fumigant. They are liposoluble and as such lodge in the primitive nervous systems of nematodes, and kill primarily through physical rather than chemical action. These nematicides had 1% to 2% chloropicrin (tear gas) added as a warning agent.

8.3 The halogenated hydrocarbon nematicides are _____ fumigants and exert their lethal effect by _____ action.

- -

narcotic   physical

8-B ORGANOPHOSPHATES

Because nematodes have nervous systems similar to those of insects (though more primitive), they are susceptible to the action of the organophosphate insecticides. Unfortunately, most of the organophosphates commonly used as insecticides are degraded rapidly in the soil and only the systemics are effective as nematicides. As a result, we find only a few of the insecticidal materials involved, several of which have been developed especially for their nematicidal effects. Terbufos (Counter[R]) is an excellent example (see OP insecticides). Others that have fallen by the wayside are dichlofenthion (Nemacide[R], Mobilawn[R]), thionazin (Zinophos[R]) and diamidfos (Nellite[R]), whose structures are not shown. Those organophosphates which are used with success are phorate, disulfoton, ethoprop, fensulfothion, and fenamiphos.

PHORATE (Thimet®)

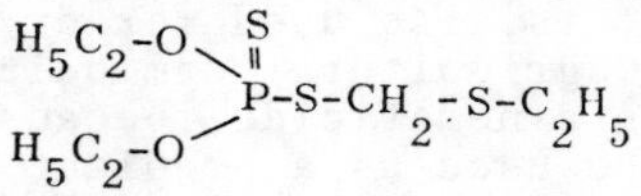

$$H_5C_2-O \diagdown \overset{\overset{\text{S}}{\|}}{P}-S-CH_2-S-C_2H_5$$
$$H_5C_2-O \diagup$$

0,0-diethyl S-(ethylthio)methyl phosphorodithioate

DISULFOTON (Disyston®)

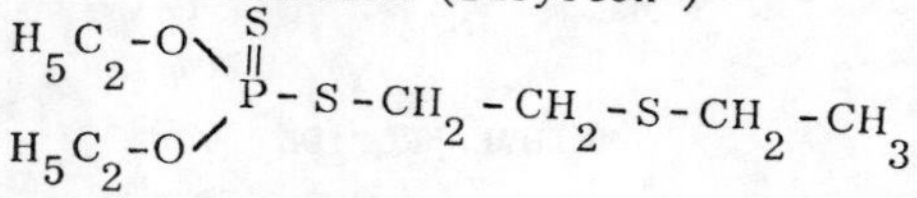

$$H_5C_2-O \diagdown \overset{\overset{\text{S}}{\|}}{P}-S-CH_2-CH_2-S-CH_2-CH_3$$
$$H_5C_2-O \diagup$$

0,0-diethyl S(2-(ethylthio)ethyl) phosphorodithioate

ETHOPROP (Mocap®)

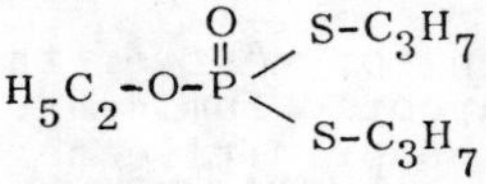

$$H_5C_2-O-\overset{\overset{\text{O}}{\|}}{P} \diagdown \overset{S-C_3H_7}{\underset{S-C_3H_7}{}}$$

0-ethyl S,S-dipropyl phosphorodithioate

FENSULFOTHION (Dasanit®)

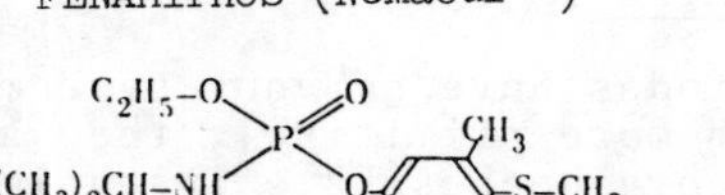

$$H_5C_2-O \diagdown \overset{\overset{\text{S}}{\|}}{P}-O- \bigcirc -\overset{\overset{\text{O}}{\|}}{S}-CH_3$$
$$H_5C_2-O \diagup$$

0,0-diethyl-0-[p-(methylsulfinyl)phenyl] phosphorothioate

FENAMIPHOS (Nemacur ^{R})

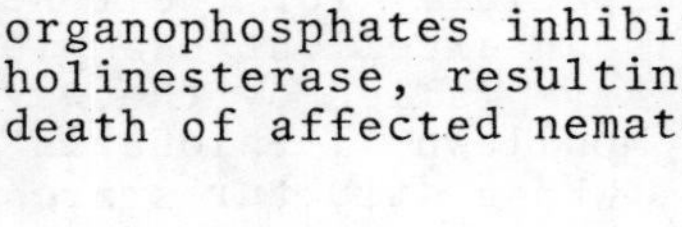

$$C_2H_5-O \diagdown \overset{\overset{\text{O}}{\|}}{P} \diagup O- \bigcirc \overset{CH_3}{\underset{S-CH_3}{}}$$
$$(CH_3)_2CH-NH \diagup$$

ethyl 3-methyl-4-(methylthio)-
phenyl (1-methylethyl)phosphoramidate

All of the organophosphates inhibit the neurotransmitter enzyme, cholinesterase, resulting in paralysis and
utlimately, the death of affected nematodes.

8.4   The mode of action of the organophosphate nematicides
      is that of inhibiting the enzyme _____.
----
      cholinesterase

8-C   ISOTHIOCYANATES

      There are three nematicides in this classification:
Metam-sodium, Vorlex^R, and dazomet.   Metam-sodium is a
dithiocarbamate mentioned under the fungicides, but it is
readily converted to an isothiocyanate, and is active
against all living matter in the soil: Weeds, weed seeds,
nematodes and soil fungi.

METAM-SODIUM (SMDC, Vapam^R)

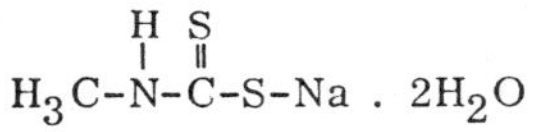

$$H_3C-\overset{\overset{\displaystyle H}{|}}{N}-\overset{\overset{\displaystyle S}{\|}}{C}-S-Na \: . \: 2H_2O$$

sodium N-methyldithiocarbamate dihydrate

      Vorlex^R also discussed with the fungicides, is a pre-
plant fumigant effective against nematodes, fungi, and
weeds.

Vorlex^R

$$CH_3-N=C=S$$

methylisothiocyanate

      Dazomet is a diazine, resembling the thiazoles
slightly, and undergoing a similar ring cleavage in the
soil to produce methylisothiocyanate.

DAZOMET (DMTT)

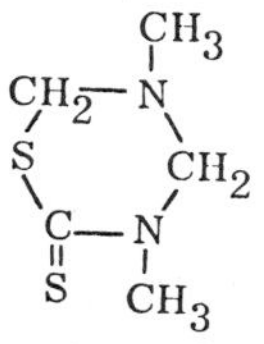

tetrahydro-3,5-dimethyl-2H,1,3,5-thiadiazine-2-thione

8.5   We also recall that the isothiocyanates act by inac-
      tivating the _____ groups in enzymes.

      -SH

8-D CARBAMATES

Aldicarb is the only example of an oxime nematicide, and was mentioned earlier in the section on carbamate insecticides as a systemic. It is registered on several crops, for example, cotton, potatoes, sugar beets, sorghum, oranges, pecans, peanuts, sweet potatoes, and ornamentals, to name a few. Unlike most of the other nematicides, aldicarb is formulated only as the granular material because of its high acute toxicity. It is drilled into the soil at planting, or after the plants are in various stages of growth, becomes water-soluble, and is absorbed by the roots and translocated throughout the plant. Its promise for expanded use is reduced because it has been found in ground water supplies in areas where it was extensively used over long periods.

ALDICARB (Temik^R)

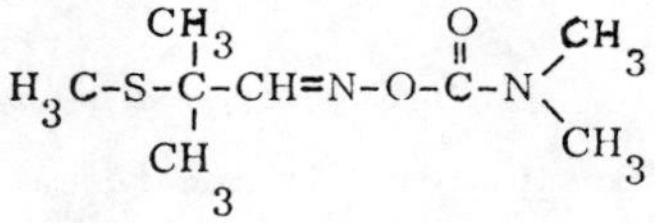

$$H_3C-S-\underset{\underset{CH_3}{|}}{\overset{\overset{CH_3}{|}}{C}}-CH=N-O-\overset{\overset{O}{\|}}{C}-N\underset{CH_3}{\overset{CH_3}{<}}$$

2-methyl-2-(methylthio)propionaldehyde O-(methylcarbamoyl) oxime

Carbofuran is a prominent systemic carbamate insecticide/nematicide. It is currently registered as a nematicide for field corn, tobacco, peanuts, sugarcane, soybeans, grapes and cucurbits. It has a fairly short residual life, making it useful on forage and vegetable crops.

CARBOFURAN (Furadan^R)

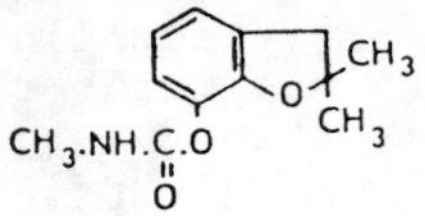

$$CH_3.NH.\overset{\overset{O}{\|}}{C}.O$$

2,3-dihydro-2,2-dimethyl-7-benzofuranyl methylcarbamate

Oxamyl is a carbamate insecticide, nematicide and acaricide, useful on many field crops, vegetables, fruits and ornamentals. It is totally soluble in water, accounting for its effectiveness as a nematicide. Carbosulfan is an insecticide/nematicide/acaricide used on alfalfa, citrus, corn, deciduous fruit, sorghum, soybeans, and other vegetable, field, and orchard crops, and may be applied to foliage or soil.

OXAMYL (Vydate^R)

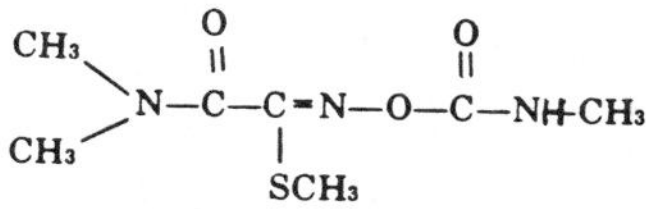

methyl N',N'-dimethyl-N-[(methyl carbamoyl)
oxy]-1-thiooxamimidate

CARBOSULFAN (Advantage^R)

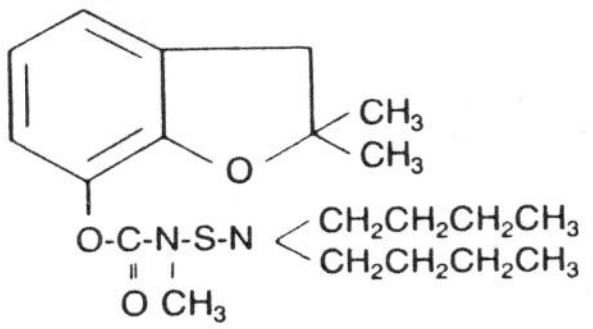

2,3-dihydro-2,2-dimethyl-7-benzofuranyl-
[(dibutylamino)thio] methyl carbamate

8.6 From memory (or by turning back to the insecticides),
the carbamates' mode of action is the inhibition of
the enzyme _____.

- -

cholinesterase (ChE)

With nematicides in general, the residue problem becomes serious with the EPA examining more closely the chlorinated and brominated fumigants for subtle, long-range
health effects. Their use will undoubtedly decline as long-
term feeding studies on laboratory animals reveal untoward
results.
Consequently, the nonvolatile carbamates and organophosphates, with their short residual effects, are the
nematicides of the future.

8.7 Why will the volatile, traditional nematicides gradually decline and be replaced by the OPs and carbamates?

- -

EPA is phasing out chlorinated and brominated
materials.

UNIT 9
RODENTICIDES

There are many kinds of small mammals, especially rodents, which damage man's dwellings, his stored products, and his cultivated crops. Among these are native rats and mice, squirrels, woodchucks, pocket gophers, hares, and rabbits. Rats are notorious freeloaders, and in some of the underprivileged countries where it is necessary to store grain in the open, as much as 20% may be consumed by rats before man has access to it.

The order Rodentia comprises about one-half of all mammalian species, and, because they are so very highly productive and widespread, they are continuously competing with man for his food. Most of the methods used to control rodents are aimed at destroying these small mammals. Poisoning, shooting, trapping, and fumigation are among the methods selected. Of these poisoning is most widely used and probably the most effective and economical. Because rodent control is in itself a diverse and complicated subject, we will mention only those more commonly used rodenticides.

The rodenticides differ widely in their chemical nature. Strange to say, they also differ widely in the hazard they present under practical conditions, even though all of them are used to kill animals that are physiologically similar to man.

9.1 Which order of animals comprise about one-half of all mammalian species?

_____

- -

Rodentia

9.2 What are the 4 most common methods of controlling rodents?

_____ _____ _____ _____

- -

poisoning, shooting, trapping, and fumigation

9.3   The most effective and economical means of controlling
      rodents is by _____?

- -

      poisoning

9-A   PHOSPHORUS

      Elemental phosphorus occurs in two common forms, the
relatively harmless red and the highly toxic white or
yellow phosphorus. Phosphorus is used but little today,
having been replaced by the anticoagulants. Yellow phosphorus attacks the liver, kidney, and heart, resulting in
rapid tissue disintegration; it also causes rats to attempt
to vomit, a function which uniquely they cannot perform.
      Zinc phosphide (Zn_3P_2), is an intense poison to mam-
mals and birds. Against rats, mice, gophers, ground squirrels, and prairie dogs, it is still used in large quantities. Its mode of action is similar to that of phosphorus.

9-B   COUMARINS (ANTICOAGULANTS)

      There are seven compounds belonging to this classifi-
cation, all of which have been very successful rodenticides.
Their mode of action is twofold: (1) inhibition of pro-
thrombin formation, the material in blood responsible for
clotting, and (2) capillary damage, resulting in internal
bleeding. The earlier coumarins require repeated ingestion over a period of several days, leaving the unsuspect-
ing rodents growing weaker by the day. The earlier coumarins are thus considered relatively safe, since repeated
accidental ingestion would be required to produce serious
illness. In the case of most other rodenticides a single
accidental ingestion could be fatal. Of the available
rodenticides, only the earlier anticoagulants are safe for
home use. Vitamin K_1 is the antidote for accidental poisoning by these and all other anticoagulant rodenticides.

9.4   What are the two effects of coumarins on rats?

      _____   _____

- -

      (1) prevent blood clotting
      (2) cause internal bleeding

      Dicumarol was the first coumarin, introduced in 1948,
and was hit upon after identifying that molecule as the
compound responsible for sweet clover's toxicity to cattle.
As a rodenticide, however, dicumarol was superseded by

Warfarin which was released in 1950 by the Wisconsin Alumni
Research Foundation (thus its name). It was immediately
successful as a rat poison because rats did not develop
"bait shyness" as they did with other baits during the required ingestion period of several days.

9.5 Why are the coumarins the safest rodenticides used?

- -

Repeated doses are necessary to produce accidental
poisoning

9.6 The coumarins' successful use-history was due to the
fact that rats didn't develop _____
during the poisoning period.

- -

bait shyness

Coumachlor was introduced in 1953, but was never
successful in the U. S. because of Warfarin's wide acceptance. Coumatetralyl was developed in Germany and introduced in the U. S. in 1957 with a fair degree of success
in situations where Warfarin had resulted in bait shyness
of rodents. Coumafuryl is also one of the commonly used
anticoagulants.

WARFARIN

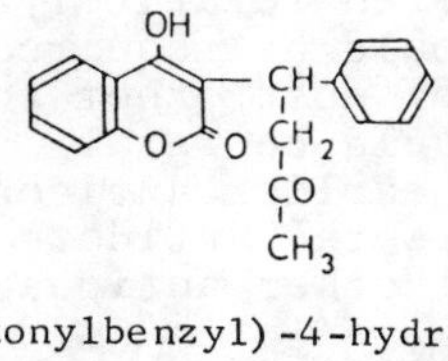

3-(α-acetonylbenzyl)-4-hydroxycoumarin

COUMAFURYL(Fumarin[R])

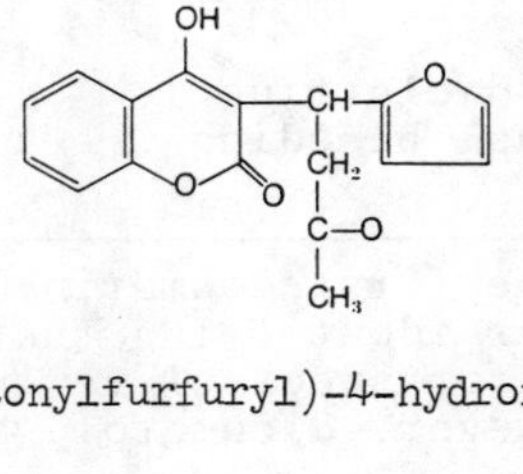

3-(a-acetonylfurfuryl)-4-hydroxycoumarin

COUMATETRALYL (Racumin[R])

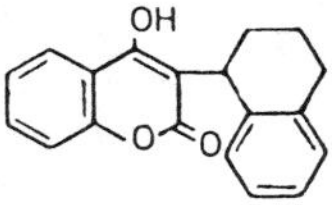

3-(d-tetralyl)-4-hydroxycoumarin

Two unique rodenticides have appeared in the past six years, brodifacoum and bromadiolone. These are different from the earlier coumarins in that, though they are anti-coagulant in their mode of action, they require but a single feeding for rodent death to occur, which requires from 4 to 7 days. They are both effective against rodents that are resistant to conventional anticoagulants. Because these two rodenticides are restricted use pesticides, they are not available to the homeowner.

BRODIFACOUM (Talon[R], Havoc[R])

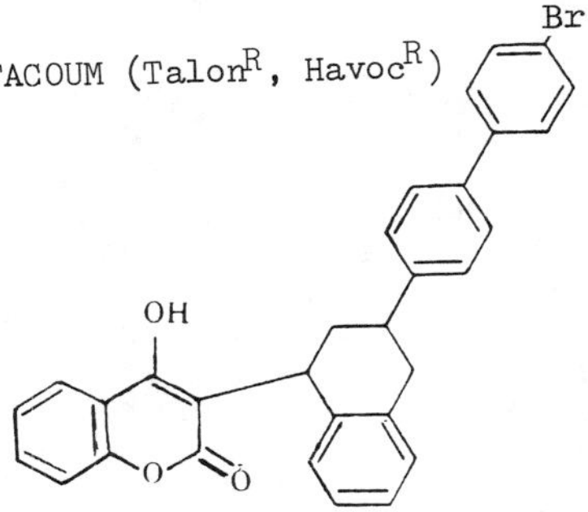

3-(3-(4'-bromo(1-1'-biphenyl)-4-yl)-
1,2,3,4-tetrahydro-1-napthalenyl)-4-
hydroxy-2H-1-benzopyran-2-one

BROMADIOLONE (Maki[R], Contrac[R])

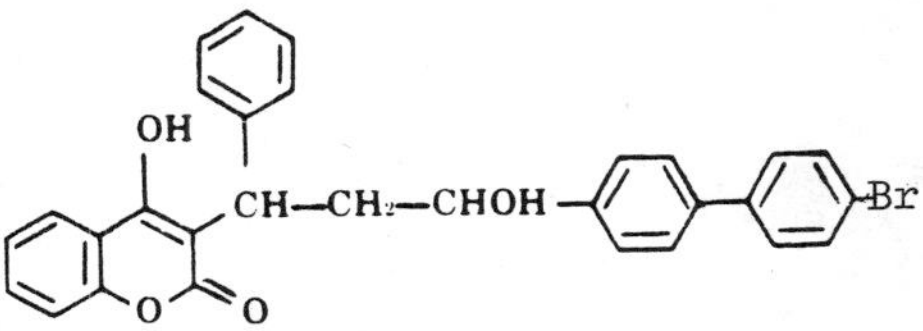

3-(3-(4'-bromo(1,1'-biphenyl)-4-yl)-
3-hydroxy-1-phenylpropyl)-4-hydroxy
2H-1-benzopyran-2-one

For protection of children and pets from accidental
ingestion of these highly toxic anticoagulant rodenticides,
baits must be placed in tamper-proof bait boxes or in locations not accessible to children.

These two rodenticides are relatively specific for
rodents by being offered as rodent-attractive, pre-mixed
baits, inaccessible to pets and domestic animals, but
especially by their relative toxicity to other warm-blooded
animals. For instance, the rat is two times more sensitive
to brodifacoum than the pig and dog, 36 times more than the
chicken, and 90 times more sensitive than the cat.

9.7 How are brodifacoum and bromadiolone different from
     the earlier coumarin rodenticides?

----

     require only one feeding

9-C INDANDIONES (ANTICOAGULANTS)

There are three compounds, pindone, diphacinone and
chlorophacinone, belonging to this class of anticoagulants
that differ chemically from the coumarin anticoagulants.
Pindone appeared in 1942, and was truly the first anticoagulant rodenticide. Daily feedings by rodents are
necessary to produce death.

Diphacinone appeared in the early 1950s, and was the
first single-dose anticoagulant. In most instances two or
three feedings are actually necessary to kill rats. Chlorophacinone was registered in 1961, and does indeed require
but a single dose of bait containing 50 mg/kg, killing rats
from the fifth day. All three of these indandiones induce
bait shyness.

Chlorophacinone, in addition to its anticoagulant
property also uncouples oxidative phosphorylation, helping
explain its success as a single-dose rodenticide. Again,
vitamin K_1 is the antidote for accidental poisoning.

PINDONE (Pival®)   DIPHACINONE (Diphacin[R], Promar[R], Ramik[R])

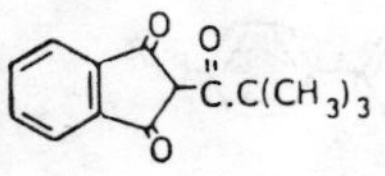

2-pivaloylindane-1,3-dione

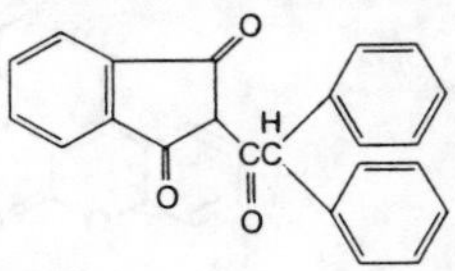

2-diphenylacetyl-1,3-indandione

CHLOROPHACINONE (Rozol^R)

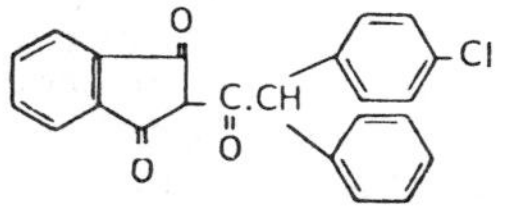

2-[(2-p-chlorophenyl)-2-phenylacetyl]-1,3-indandione

9.8 Presently used rodenticides are effective due to which two actions:

_____

_____

- -

prevent blood clotting
cause internal bleeding following capillary damage

9-D BOTANICALS

Red Squill, which comes from the powdered bulbs of a plant, Mediterranean squill, was used before 1935, but was never more than a mediocre rodenticide. The active ingredient is scilliroside, classed as a cardiac glycoside. Its specific activity is due to the inability of rats to vomit, thus they must absorb the toxicant. Other animals ingesting squill do vomit, permitting them to survive accidental poisoning.

RED SQUILL (Scilliroside)   STRYCHNINE

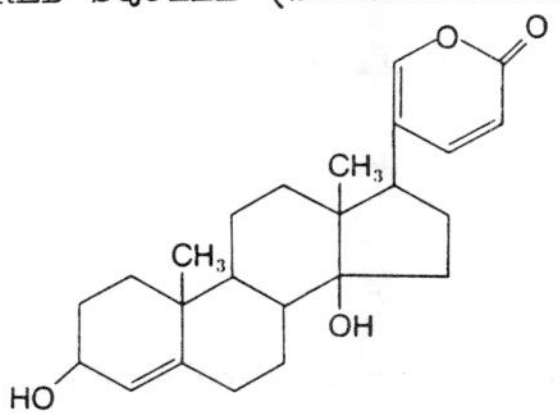

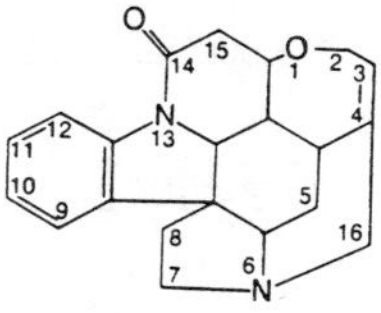

Strychnine is an alkaloid from an Asiatic tree, *Strychnos nux-vomica*, that is usually converted to strychnine sulfate for use as a rodenticide. Strychnine is highly toxic to all warm-blooded animals, and acts by paralyzing specific muscles, resulting in cessation of breathing and heart action. It is formulated as colored grain baits containing 0.5% to 1.0% strychnine sulfate.

9-E   ORGANOCHLORINES

DDT 50 percent dust was used for years by structural
pest control operators as a tracking powder. The dust was
sprinkled in the known runs of mice, and, after tracking
through the dust, the mice stopped to preen themselves and
clean their feet. Death resulted from convulsions and paralysis, just as in insects. DDT was exceptionally effective
against bats.

Endrin is still registered for application to orchard
soils and fruit tree trunks during the fall or winter
months, for vole control. These field mice, while eating
the bark and trailing across treated soil, quickly ingest
lethal quantities. (See structures in section on organochlorine insecticides.)

9-F   MISCELLANEOUS RODENTICIDES

Compound 1080. Sodium fluoroacetate, one of the most
toxic poisons known, was introduced in 1947. Its use is now
highly restricted to authorized trained personnel because
of its extreme hazard to man and domestic animals. Sodium
fluoroacetate has a strong effect on both the heart and
nervous system, resulting in convulsions, paralysis, and
death.

9.9   Why can't 1080 be used by just anyone with rodent
      problems?

- -

too toxic for laymen to use

SODIUM FLUOROACETATE (1080^R)   FLUOROACETAMIDE (1081)

$$F-CH_2-\overset{\displaystyle O}{\overset{\|}{C}}-O-Na \qquad\qquad F-CH_2-\overset{\displaystyle O}{\overset{\|}{C}}-O-NH_2$$

Compound 1081 is a moderately fast-acting rodenticide
closely related to sodium fluoroacetate. It possesses a
lower mammalian toxicity and a longer latent period before
animals become distressed and stop feeding. Its use is
less likely to lead to poison shyness because of sublethal
dosing. All precautions pertaining to 1080 apply also to
1081.

Antu[R] derived its name from the chemical nomenclature given it when introduced in 1946. (See if you can find the cryptic name Antu in its chemical name.) Because rodents develop a tolerance to the material, it has been displaced by other materials.

ANTU

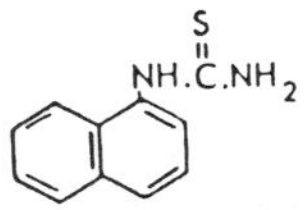

$$NH.\overset{\overset{S}{\|}}{C}.NH_2$$

α-naphthylthiourea

Thallium sulfate (TlSO4) is an old rodenticide now available only for use by government agencies, to control rats, moles and house mice. It is a general cellular toxin resembling arsenic in its effects, and attacks or inhibits enzymes other than those containing -SH groups.

Pyriminil (Vacor[R]) appeared in 1977 as an acute toxicant that killed rodents in 4 to 8 hours after ingestion of a single dose. Its mode of action is the inhibition of niacinamide metabolism, causing rodents to die from paralysis and pulmonary arrest. Apparently pyriminil was surprisingly toxic to humans and it was removed voluntarily from the market by its manufacturer in 1978 following several near fatal human exposures.

9.10 Give 3 reasons why a particular rodenticide may not be selected for use:

_____   _____

_____

- -

(1) too toxic
(2) not very effective
(3) rats develop bait shyness

UNIT 10
PLANT GROWTH REGULATORS

Though not truly pesticides in the usual sense of the word, chemicals used in some way to alter the growth of plants, blossoms or fruits, are plant growth regulators (PGRs), and legally are pesticides. Plant growth regulators are also referred to as plant regulating substances, growth regulants, plant hormones, and plant regulators.

Plants contain natural substances that control growth, initiate flowering, cause blossoms to fall, set fruit, cause fruit and leaves to fall, control initiation and termination of dormancy, and stimulate root development. These natural substances are hormones.

10.1 Name 5 functions of plant growth regulators.

    _____   _____

    _____   _____

    _____

- -

control growth, initiate flowering, cause blossoms to fall, set fruit, cause fruit and leaves to fall, control initiation and termination of dormancy, stimulate root development

---

10.2 Plant growth and development processes are controlled by _____.

- -

hormones

The history of PGRs begins in 1932, when it was discovered that acetylene and ethylene would promote flowering in pineapple. In 1934, auxins were found to enhance root formation in cuttings. Since then, outstanding developments have occurred as a consequence of the use of PGRs. A few milestones were: The development of seedless fruit; prevention of berry and leaf drop in holly; prevention of early, premature drop of fruit; promotion of heavy setting of fruit blossoms; thinning of blossoms and fruit; prevention of sprouting in stored potatoes and onions; and the

inhibition of buds in nursery stock and fruit trees to
prolong dormancy. These are just a few of the hundreds
of achievements made possible with these growth-controlling
chemicals. As a result, we eat better quality foods, eat
fruits and vegetables out-of-season or literally year-round,
and pay less for our food because of the reduced need for
hand labor in thinning and harvesting.

10.3 Of what value are plant growth regulators to the
     average person?

- -

        eat better quality, have fruits and vegetables year-
        round, pay less for our food

        The American Society for Horticultural Science recog-
nizes six classes of PGRs: auxins, gibberellins, cytokinins, ethylene generators, inhibitors, and growth retardants. These very useful agricultural chemical tools will
be discussed under these headings.

10-A AUXINS

        Auxins are compounds that induce elongation in shoot
cells. Some occur naturally whereas others are manufactured. Auxin precursors are materials that are metabolized
to auxins in the plant. Antiauxins are chemical compounds
that inhibit the action of auxins.

10.4 Auxins cause _____ in shoot cells.

- -

        elongation

        They are used to thin apples and pears, increase
yields of potatoes, soybeans, and sugar beets, to assist
in the rooting of cuttings, and to increase flower formation, among other things. The mechanism of action is not
completely understood, but they may work by controlling the
type of enzyme produced in the cell. In any event, with
the addition of auxin the individual cells become larger by
a loosening of the cell wall, which is followed by the
increased uptake of water and expansion of the cell wall.

10.5 What are 4 uses of auxins in crop production?

- -

thin apples and pears; increase yields of potatoes,
soybeans, and sugar beets; to assist in the rooting
of cuttings; to increase flower formation

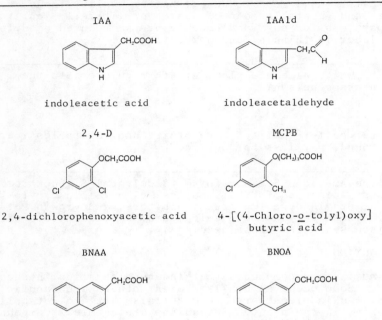

IAA	IAA1d
indoleacetic acid	indoleacetaldehyde
2,4-D	MCPB
2,4-dichlorophenoxyacetic acid	4-[(4-Chloro-o-tolyl)oxy] butyric acid
BNAA	BNOA
β-naphthaleneacetic acid	β-naphthoxyacetic acid

    2,4-D is classed as a herbicide with auxin-like char-
acteristics. In citrus culture, 2,4-D is used to prevent
preharvest fruit–drop in older trees (6 years and older),
to prevent leaf- and fruit–drop following pesticide oil
sprays, to delay fruit maturity, and to increase fruit
size. To achieve these effects 2,4-D is applied at precise
times, to the entire tree, as water sprays containing the
herbicide at 8 to 16 ppm. It is effective on grapefruit,
lemons, and navel and Valencia oranges.

10-B GIBBERELLINS

    Gibberellins are compounds having a *gibbane* skeleton;
they stimulate cell division or cell elongation or both.
In 1957, gibberellin, or gibberellic acid, was introduced
to the horticultural world. It caused fantastic growth in
many types of plants, and though originally isolated from
a fungus, gibberellins were later found to be natural constituents in all plants. Since then, more than 57 gibberellins have been isolated and are identified as GA_1, GA_2,
etc. The gibberellic acid most commonly used is GA_3.

The mechanism of action for gibberellin is the induction
or manufacture of more enzyme(s) in the cells, resulting
in cell growth, particularly by elongation. The most
striking effect of treating with gibberellin is the stimu-
lation of growth, expressed as long stems.

10.6 Gibberellins cause more _____ to be produced
     in cells resulting in cell growth by _____ .

----

     enzymes   elongation

     For example: gibberellins are used to increase stalk
length and yields of celery, to break dormancy of seed
potatoes, to increase grape size, to induce seedlessness in
grapes, to improve the size of greenhouse-grown flowers,
to delay fruit maturity, to extend fruit harvest, and to
improve fruit quality.

GIBBANE GA_3

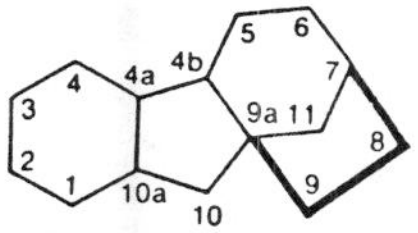

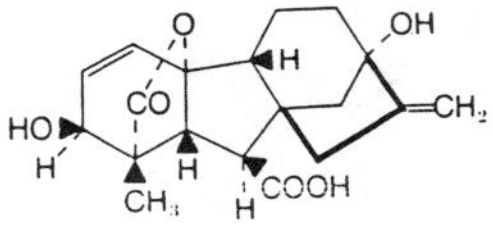

                                  gibberellic acid

10.7 List 5 uses of gibberellins:

----

     to increase stalk length and yields of celery, to
     break dormancy of seed potatoes, to increase grape
     size, to induce seedlessness in grapes, to improve
     the size of greenhouse-grown flowers, to delay fruit
     maturity, to extend fruit harvest, to improve fruit
     quality

10-C CYTOKININS (PHYTOKININS)

     Cytokinins, sometimes referred to as phytokinins, are
naturally occurring or manufactured compounds that induce
cell division in plants. Most of the cytokinins are derivatives of adenine. These useful materials were discovered
in 1955, and their practical potential lies in prolonging
the storage life of green vegetables, cut flowers, and
mushrooms.
     The cytokinins cause two outstanding effects — the
induction of cell division and the regulation of differen-
tiation in removed plant parts. Their mechanism of action

is not known, but they apparently act at the gene level, becoming incorporated in the nucleic acids of the cell. This ultimately influences cell division, in which the nucleic acids play critical roles.

10.8 Cytokinins are currently used to prolong ________ of fresh produce.

----

storage

NATURALLY OCCURRING CYTOKININS

ZEATIN

CH₃
HNCH₂—CH═C—CH₂OH
N
N
N
N
H

2iP

CH₃
HNCH₂—CH═C—CH₃
N
N
N
N
H

6-(γ,γ-dimethylallylamino) purine

10.9 List 3 uses of cytokinins:

----

prolonging the storage life of green vegetables, cut flowers, and mushrooms

---

SYNTHETIC CYTOKININS

KINETIN

NH—CH₂
O
N
N
N
N
H

6-furfurylamino purine

BA

NHCH₂
N
N
N
N
H

6-Benzylamino purine

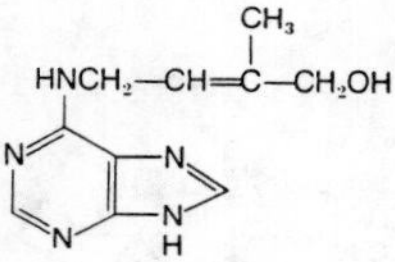

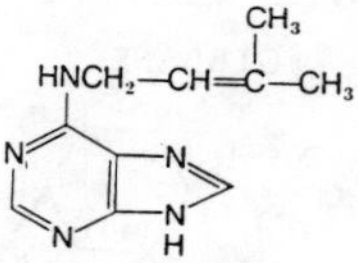

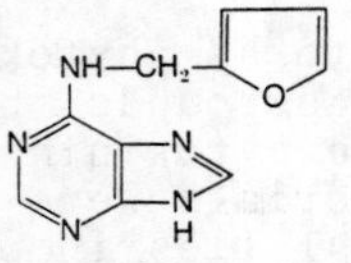

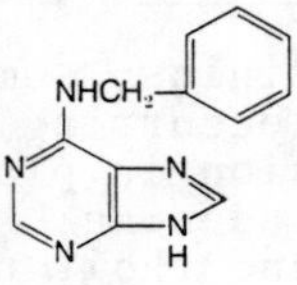

ADENINE

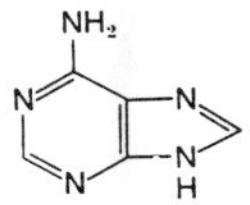

PBA

6-aminopurine

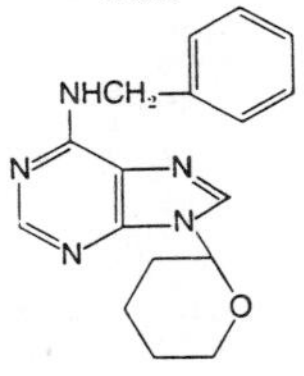

6-(Benzylamino)-9-(2-tetrahydro-
pyranyl)-9H-purine

10-D ETHYLENE GENERATORS

Recall that the first mention of a plant growth regulator was the use of ethylene in promotion of pineapple flowering. Recently, materials have been developed that are applied as sprays to the growth sites of plants and that stimulate the release of ethylene ($H_2C=CH_2$). Ethylene produces numerous physiological effects and can be used to regulate different phases of plant metabolism, growth and development.

As with the other growth regulating chemicals, the ethylene generators have many uses. They can be used to accelerate pineapple maturity; induce uniform fig ripening; facilitate mechanical harvest of peppers, cherries, plums, and apples; induce uniform flowering and fruit thinning, among other possibilities.

10.10 Name 4 uses of ethylene generators in food production.

- -

accelerate pineapple maturity; induce uniform fig ripening; facilitate mechanical harvest of peppers, cherries, plums and apples; induce uniform flowering and fruit thinning

---

When introduced at the proper time, ethylene can be used to stimulate seed germination and sprouting, abscission of flowers, leaves and fruit, regulation of growth and ripening of fruit. None of these biological effects results from a clearly defined mechanism of action. Generally, however, ethylene apparently has its greatest influence on the dominant enzymes of that particular physiological state of the absorbing tissue. The ethylene apparently serves as a trigger or synergist, resulting in a chain of premature biochemical events, expressed in the final and usually desirable result. Ethephon gradually releases ethylene as a degradation product when applied to plant surfaces.

ETHEPHON (Ethrel^R, Florel^R, Prep^R)

10.11 Ethylene generators can have _____
      effects on plants, depending on their physiological
      state.

- -

      multiple or many

10-E INHIBITORS AND RETARDANTS

Inhibitors are an assorted group of substances that
inhibit a physiological process in plants. Those that
occur naturally in plants are usually hormones, and inhibit
different functions; for instance, growth or seed germination, or the action of other hormones, gibberellins, and
auxins. New types of synthetic compounds, plant growth
retardants, have recently been discovered.

The inhibitors and retardants have many uses: they
prevent sprouting of stored onions, potatoes and root
crops, they retard sucker development on tobacco plants,
they induce shortening of stems, they favor redistribution
of dry matter, they prevent lodging of grain, and they permit controlled growth of flower crops and ornamentals.

These materials are a diverse group of compounds and
thus have different biological effects on plants. The
inhibitors and retardants are antagonistic to the growth-
promoting hormones such as auxins, gibberellins and cytokinins, through a potential multitude of biochemical
actions. We leave the plant growth regulators with this
veil of mystery surrounding their curious yet beneficial
effects.

10.12 The inhibitors and retardants are antagonistic to
      the growth _____ .

- -

      hormones

NATURALLY OCCURRING INHIBITORS

BENZOIC ACID

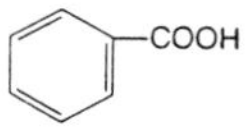

GALLIC ACID

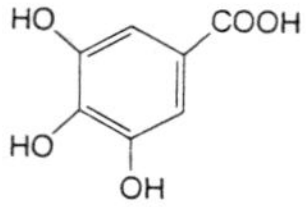

CINNAMIC ACID

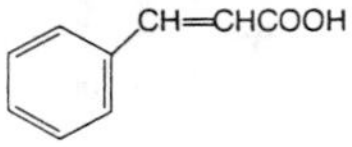

(S)-ABSCISIC ACID

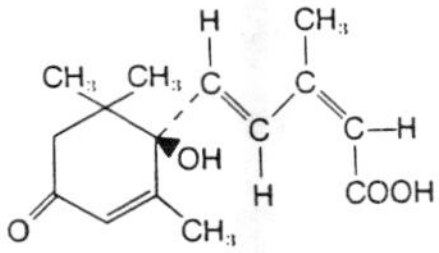

2-TRANS-ABSCISIC ACID

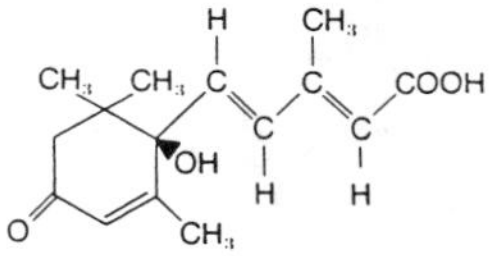

10.13 List 4 uses of inhibitors and retardants:

----

prevent sprouting of stored onions, potatoes and root crops; retard sucker development on tobacco plants; induce shortening of stems; favor redistribution of dry matter; prevent lodging of grain; permit controlled growth of flower crops and ornamentals

---

SYNTHETIC INHIBITORS

MH - MALEIC HYDRAZIDE

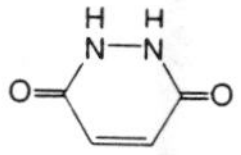

CHLORMEQUAT CHLORIDE (Cycocel[R])

$$ClCH_2CH_2N^+(CH_3)_3Cl^-$$

2-chloroethyltrimethylammonium ion

1,2-dihydro-3,6-pyridazinedione

Chlormequat chloride has a wide range of uses, especially in the production of flowers by shortening internodes and thickening stems. Maleic hydrazide is a growth retardant used for the short-term growth inhibition of trees, shrubs and grasses, and also for controlling the sprouting of onions and potatoes in storage. It has many other uses.

Mepiquat-chloride (Pix[R]) is a growth regulator for cotton. When applied at early bloom it reduces vegetative growth. In some instances it is reported to increase cotton maturity and yields. Mepiquat-chloride in combination with ethephon is used as a growth regulator (Terpal[R]) for winter and spring barleys, rye and oats, to speed heading and early harvest. The mode of action for mepiquat-chloride is as yet unknown.

MEPIQUAT CHLORIDE (Pix[R])

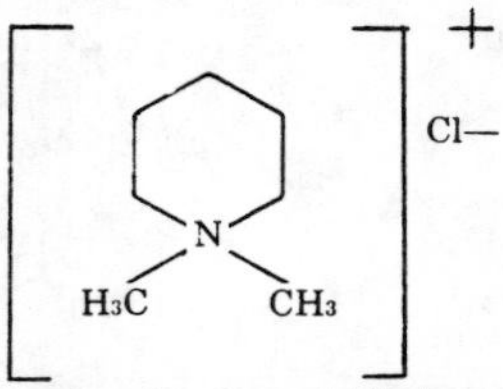

1,1-dimethyl-piperidiniumchloride

Ancymidol reduces internode elongation and is used in greenhouse flower production applied either to soil or foliage. Daminozide had all registrations cancelled by EPA in September 1985, because it was found to be a carcinogen in laboratory animals. It was widely used on peanuts, apples and other fruits. Dikegulac serves as a pinching agent for azaleas and as a growth retardant for shrubs. Mefluidide suppresses seedhead formation and regulates growth of turf grasses. Additionally, it is registered as a growth regulator to improve the effects of certain herbicides.

Tecnazene (Fusarex[R]), structure not shown, is both a fungicide and growth regulator. It is used primarily to control dry rot and inhibit sprouting in stored potatoes. Ethoxyquin (Deccoquin 305[R]), also not shown, is applied to deciduous fruit during development to reduce fruit scald.

ANCYMIDOL (A-Rest[R])

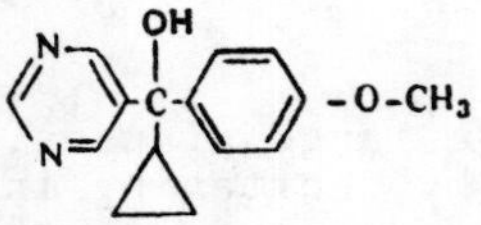

a-cyclopropyl-a-(p-methoxyphenyl)-
5-pyrimidine methanol

DAMINOZIDE (Alar[R])

$$(CH_3)_2N-NHCCH_2CH_2COH$$

butanedioic acid mono-
(2,2-dimethyl hydrazide)

DIKEGULAC SODIUM (Atrinal^R)

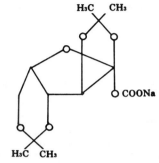

sodium salt of 2,3:4,6-bis-0
(1-methylethylidene)-0-
(-L-xylo-2-hexulofuranosonic acid)

MEFLUIDIDE (Embark^R)

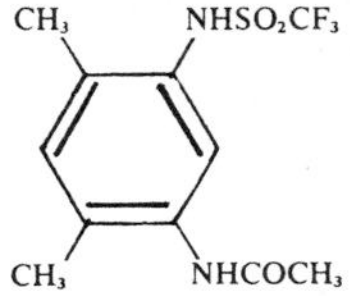

N-(2,4-dimethyl-5-(((trifluoromethyl)
sulfonyl)amino)phenyl)acetamide

10.14 Now, list the 6 classes of plant growth regulators:

----

auxins, gibberellins, cytokinins, ethylene generators,
inhibitors, and retardants

UNIT 11
DEFOLIANTS

Defoliants are chemicals that accelerate leaf fall from crop plants such as cotton, soybeans, or tomatoes, and thereby facilitate harvest procedures. Let's use the defoliation of cotton as an example. The premature removal of leaves from the cotton plant permits earlier harvesting, and results in higher grades of cotton because few leaves remain to clog the mechanical picker, add trash, or stain the fibers. Defoliation often helps lodged (fallen-over) plants to straighten up, increasing the plant exposure to sun and air. This enables the plants to dry quickly and thoroughly and the mature bolls open faster to reduce boll rots that damage fiber and seed. And finally, defoliation of the cotton plant reduces populations of fiber-staining insects, particularly aphids and whiteflies, which deposit honeydew in the open bolls.

11.1 The function of defoliants is to _____ _____ _____ which _____ _____ .

- -

accelerate leaf fall   aids harvest

Two conditions must exist before defoliation of any crop is effective: First the plant must be in a state of maturity in which growth has stopped, and second, the temperatures must exceed 80° during the day and 50° at night.

11.2 Defoliation requires plant _____ and high _____ to be effective.

- -

maturity   temperatures

Chemical defoliation is the premature abscission of leaves, brought on by the formation of the abscission layer at the point where the leaf petiole joins the plant stem.

11.3 Abscission is the _____ of the leaf
     from the plant.

- -

     separation

11-A INORGANIC SALTS

Sodium chlorate ($NaClO_3$), magnesium chlorate
($Mg(ClO_3)_2 . 6H_2O$), disodium octaborate tetrahydrate
($Na_2B_8O_{14} . 4H_2O$), and the other sodium polyborates are
all old defoliants, still in use, and primarily on cotton.
These are contact materials, which, by virtue of their
high acidity, bring about rapid destruction of the delicate
protoplasmic structures, resulting in the formation of the
petiole abscission layer. The exact nature and sequence of
the chemical reactions is unknown. The chlorates cause
chlorosis of leaves and a starch-depletion in stems and
roots when applied in less than lethal doses.

11.4 How do the inorganic salts bring on defoliation?

- -

     high acidity causes destruction of protoplasmic
     structures and forms abscission layer

11-B ORGANOPHOSPHATES

DEF[R] and merphos are two organophosphate defoliants
that have proved extremely useful in cotton production.
Neither of these compounds is hormone-acting. Rather, they
induce abscission by injuring the leaf, causing changes in
the levels of naturally occurring plant hormones which
induce the early formation of the leaf abscission layer.
Defoliation follows treatment by 4 to 7 days.

DEF[R]   MERPHOS (Folex[R])

$(C_4H_9S)_3P=O$ $(C_4H_9S)_3P$

S,S,S-tributyl phosphorotrithioate   tributyl phosphorotrithioite

11.5 The organophosphate defoliants cause changes in
     plant _____ which result in the development
     of the leaf _____ layer.

- -

     hormones   abscission

11-C BIPYRIDYLIUMS

Paraquat, also studied in the herbicides under the
bipyridylium class, damages plant tissue very rapidly.
Its swift action results from the breakdown of plant cells
responsible for photosynthesis, giving the leaves a water-
logged appearance within a few hours of treatment. Paraquat results in the formation of OH- radicals or hydrogen
peroxide (H_2O_2) as the primary toxicants. Because most
leaves drop off, paraquat is considered a defoliant.

11.6 What is the mode of action of paraquat?

- -

     breaks down cells that carry on photosynthesis

PARAQUAT

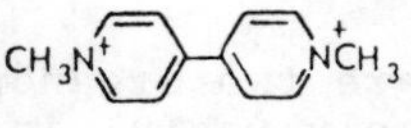

$$CH_3N^+ \diagdown \diagdown N^+CH_3$$

1,1'-dimethyl-4,4'-bipyridylium ion

11.7 The primary toxicants formed in plants by paraquat
     are _____ radicals and _____.

- -

     OH- H_2O_2

11-D MISCELLANEOUS DEFOLIANTS

Thidiazuron is a cotton defoliant. It stimulates the formation of the abscission layer causing shedding of the leaves, and also inhibits leaf regrowth.

Dimethipin is an effective cotton defoliant even at temperatures of $70°F$ and below, also for nursery stock and grapes. It speeds maturing and reduces seed moisture in rice and sunflower.

THIDIAZURON (Dropp[R])

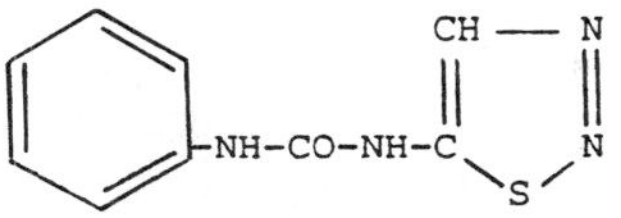

N-phenyl-N'-1,2,3-thiadiazol-5-ylurea

DIMETHIPIN (Harvade[R])

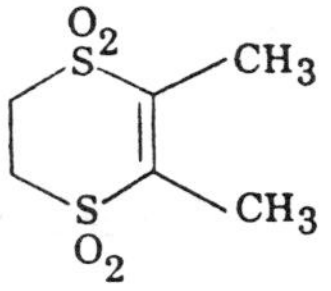

2,3-dihydro-5,6-dimethyl-
1,4-dithiin 1,1,4,4-tetraoxide

UNIT 12
DESICCANTS

Desiccants are chemicals designed to speed the drying of crop plant parts such as cotton leaves and potato vines. They usually kill the leaves rapidly, freezing them to the plant, mimicking the effect of a light frost. Desiccants cause foliage to lose water. Leaves, stems, and even branches of plants are sometimes killed so rapidly by desiccants that an abscission layer has insufficient time to develop, and the drying leaves remain attached to the plant. The advantage of desiccants over defoliants is that they can be applied at a later date than defoliants. Thus additional time is gained, during which the leaves continue to function and contribute to seed and fiber quality, at least with cotton.

12.1 What is the basic difference between defoliants and desiccants?

- -

with desiccants, leaves remain on plants

12-A THE INORGANICS

Two of the inorganic salts mentioned in earlier chapters are used as desiccants for cotton, sodium borate(s), and sodium chlorate. However, arsenic acid is probably the oldest of the cotton desiccants, and it is still used in quantity. The material penetrates the leaf cuticle and the rapid contact injury precludes any extensive translocation. Arsenic acid uncouples oxidative phosphorylation and forms complexes with SH-containing enzymes.

ARSENIC ACID

$$H_3AsO_4$$

12-B PHENOL DERIVATIVES

Dinoseb and pentachlorophenol, both phenol derivative herbicides of universal effectiveness, were also used as desiccants for cotton. All uses of pentachlorophenol as a desiccant have been cancelled by EPA.

PCP, PENTA

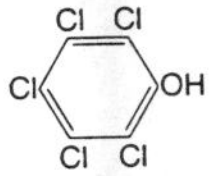

pentachlorophenol

DINOSEB, DINITRO, DNBP

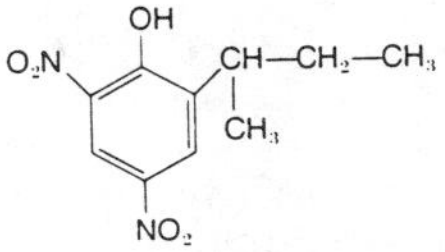

2-sec-butyl-4,6-dinitrophenol

12-C BIPYRIDYLIUMS

The bipyridylium herbicides, diquat dibromide and paraquat dichloride because of their frostlike effect on green foliage, are outstanding desiccants. Diquat dibromide is used on the seed crops only of alfalfa, clover, soybeans, and vetch, while paraquat is used on cotton, potatoes, and soybeans. (See the section on Herbicides for chemical structures.)

12-D MISCELLANEOUS DESICCANTS

Endothall, examined briefly toward the end of the chapter on herbicides, is also a desiccant. It is used on cotton and — for the seed crops only — alfalfa, clovers, soybeans, trefoil, and vetch. It kills the leaves by contact through rapid penetration of the cuticle and results in desiccation and browning of the foliage. Its mode of action is not understood.

ENDOTHALL (Accelerate^R)

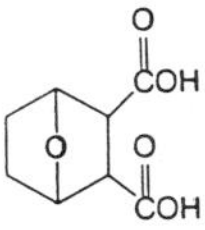

7-oxabicyclo(2,2,1)heptane-2,3-dicarboxylic acid

Ametryn, one of the very old and popular triazine herbicides, is also used as a potato vine desiccant prior to digging the potatoes. Ametryn penetrates the leaves rapidly and is a strong inhibitor of photosynthesis, causing the leaves to desiccate within 72 hours after application.

AMETRYN (Evik[R])

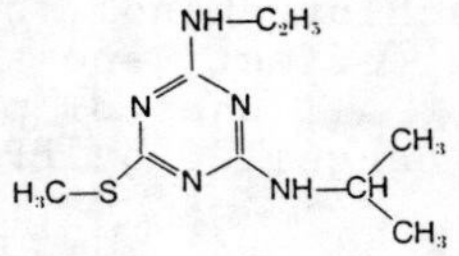

2-(ethylamino)-4-(isopropylamino)-
6-(methylthio)-S-triazine

12.2 Considering arsenic acid and dinoseb together, how do
these desiccants act biochemically?

- -

uncouple oxidative phosphorylation

12.3 Why are desiccants used instead of defoliants on
occasion?

- -

they can be applied at a later date than defoliants

UNIT 13
PESTICIDES AND THE LAW

Several federal laws protect the user of pesticides, his pets and domestic animals, wildlife, his neighbor, and the consumer of treated products. Nothing is left unprotected, and after reading briefly about the laws themselves, you will be convinced of their omnipotence.

In the beginning, the Federal Food, Drug and Cosmetic Act of 1906, known as the Pure Food Law, required that food (fresh, canned and frozen) shipped in interstate commerce be pure and wholesome. There was nothing in the law pertaining to pesticides.

In 1910, President Wm. Howard Taft signed into law the Federal Insecticide Act, which covered only insecticides and fungicides. This Act was the first to control pesticides, and was designed mainly to protect the farmer from substandard or fraudulent products, for they were abundant. It was one of our earliest laws aimed at consumer protection.

13.1 What was the first Federal law to control pesticides?

Federal Insecticide Act of 1910

In 1938, the Pure Food Law of 1906 was amended to include pesticides on foods, primarily the arsenicals such as lead arsenate and Paris green. It also required the adding of color to white insecticides, including sodium fluoride and lead arsenate, to prevent their use as flour or other look-alike cooking materials. This was the first federal effort toward protecting the consumer from pesticide-contaminated food, by providing tolerances for pesticide residues, namely arsenic and lead, in foods where these materials were necessary for the production of a food supply.

13.2 When and under what law were pesticide residues first controlled on foods?

1938   amended Pure Food Law

In 1947, the Federal Insecticide, Fungicide, and
Rodenticide Act (FIFRA) became law; it superseded the 1910
Federal Insecticide Act and extended the coverage to
include herbicides and rodenticides. It required that any
of these products must be registered with the U. S. Department of Agriculture before they could be marketed in interstate commerce. Basically, the law was one requiring good
and useful labeling, making the product safe to use if
label instructions were followed. The label was required
to contain the manufacturer's name and address, name, brand
and trademark of the product, its net contents, an ingredient statement, an appropriate warning statement to prevent
injury to man, animals, plants, and useful invertebrates,
and directions for use adequate to protect the user and the
public.

13.3 What was the first Federal pesticide labeling law?

- -

     FIFRA of 1947

In 1954, the Food, Drug and Cosmetic Act (1906, 1938)
was modified by the passing of the Miller Amendment. It
provided that any raw agricultural commodity may be con-
demned as adulterated if it contains any pesticide chemical
whose safety has not been formally cleared or that is
present in excessive amounts (above tolerances). In
essence, this clearly set tolerances on all pesticides in
food products, for example 10 ppm carbaryl in lettuce,
or 1.0 ppm ethyl parathion on string beans.

13.4 What law controls the legal pesticide residues
(tolerances) which may appear in and on foods?

- -

     1954 Miller Amendment of Food, Drug and Cosmetic Act

These two basic federal statutes, the Federal Insecticide, Fungicide and Rodenticide Act (FIFRA) and the Miller
Amendment to the Food, Drug, and Cosmetic Act, supplement
each other and are interrelated by law in practical operation. Today they serve as the basic elements of protection
for the applicator, the consumer of treated products, and
the environment, as modified by the following amendments.
In 1958, the Food Additives Amendment to the Food,
Drug and Cosmetic Act (1906, 1938, 1954) was passed. It
extended the same philosophy to all types of food additives
that had been applied to pesticide residues on raw agricultural commodities by the 1954 Miller amendment. Of greater
importance, however, was the inclusion of the Delaney

clause, which states that any chemical found to cause cancer (a carcinogen) in laboratory animals when administered
in appropriate tests may not appear in foods consumed by
humans. This has become a most controversial segment of
the spectrum of federal laws applying to pesticides, mainly
with regard to the dosage found to produce cancer in experimental animals.

13.5 Chemicals known to be carcinogens cannot appear as
     residues in foods consumed by humans. When was this
     law adopted and how is it identified?

- -

     1958   the Delaney clause of the Food Additives
            Amendment

     The various statutes mentioned so far apply only to
commodities shipped in interstate commerce. In 1959, FIFRA
(1947) was amended to include nematicides, plant regulators,
defoliants, and desiccants as economic poisons (pesticides).
(Poisons and repellents used against amphibians, reptiles,
birds, fish, mammals, and invertebrates have since been
included as economic poisons.) Because FIFRA and the Food,
Drug and Cosmetics Act are allied, these additional economic poisons were also controlled as they pertain to residues in raw agricultural commodities.

13.6 What other pesticides were included under FIFRA in
     1959?
     _____, _____ _____,
     _____ and _____.

- -

     nematicides, plant regulators, defoliants, and
     desiccants

     In 1964, FIFRA (1947, 1959) was again amended to
require that all pesticide labels contain the Federal Registration Number. It also required caution words such as
WARNING, DANGER, CAUTION, and KEEP OUT OF REACH OF CHIL-
DREN to be included on the front label of all pesticides.
Manufacturers also had to remove safety claims from all
labels.
     Until December 1970, the administration of FIFRA had
been the responsibility of the U. S. Department of Agriculture. At that time the responsibility was transferred to
the newly-designated Environmental Protection Agency (EPA).
Simultaneously, the authority to establish pesticide tolerances was transferred from the Food and Drug Administration
(FDA) to EPA, where it remains today. The enforcement of
tolerances remains the responsibility of the FDA.

In 1972, FIFRA (1947, 1959, 1964) was revised by the most important piece of pesticide legislation of this century. THE FEDERAL ENVIRONMENTAL PESTICIDE CONTROL ACT (FEPCA), referred to as FIFRA AMENDED, 1972.

Some of the provisions of FEPCA are abstracted as follows:

1. Use of any pesticide inconsistent with the label is prohibited.
2. Deliberate violations of FEPCA by growers, applicators or dealers can result in heavy fines and/or imprisonment.
3. All pesticides will be classified into (a) *restricted use* or (b) *general use* categories. (In 1985 the EPA dropped the *general use* category and adopted the term *unclassified*, which means the same thing.)
4. Anyone applying or supervising the use of restricted use pesticides must be *certified* by the state in which he lives. This provision includes both farmers and commercial applicators.
5. Pesticide manufacturing plants must be registered and inspected by EPA.
6. States may register pesticides on a limited basis when intended for special local needs.
7. All pesticide products must be registered by EPA, whether shipped in interstate or intrastate commerce.
8. For a product to be registered the manufacturer is required to provide scientific evidence that the product, when used as directed, will (1) effectively control the pests listed on label, (2) not injure humans, crops, livestock, wildlife, or damage the total environment, and (3) not result in illegal residues in food or feed.

13.7 In a few words, what are the 8 basic provisions of FEPCA:

- -

(1) you must follow the label; (2) violations are punished; (3) pesticides classed as general or restricted use; (4) restricted use pesticides require a certified applicator; (5) pesticide plants must be registered; (6) states may register pesticides for local needs; (7) all pesticides must be registered by EPA; and (8) for registration certain data must be provided.

13.8 FEPCA classified pesticides into two categories, _____ and _____.

- -

restricted use and unclassified

13.9 FEPCA requires that anyone applying or supervising the application of a restricted use pesticide must be _____.

----

certified

EPA has the responsibility to interpret the law and implement its provisions; this is done by the agency preparing regulations. Once a regulation is duely processed, it has the force of law. Usually regulations are developed by consulting with those who will be most affected. Such was the case when the applicator training and certification programs were developed.

By regulation, the EPA established ten categories of certification for commercial applicators: (1) agricultural pest control (plant and animal); (2) forest pest control; (3) ornamental and turf pest control;(4) seed treatment; (5) aquatic pest control; (6) right-of-way pest control; (7) industrial, institutional, structural, and health-related pest control; (8) public health pest control; (9) regulatory pest control; and (10) demonstration and research pest control.

General standards of knowledge were also set for all categories of certified commercial applicators. Testing is based, among other things, on the following areas of competency: (1) label and labeling comprehension, (2) safety, (3) environment, (4) pests, (5) pesticides, (6) equipment, (7) application techniques, and (8) laws and regulations. Much of this material is covered in this book. In each state, certification is carried out by an appropriate regulatory agency (Lead Agency), often the State Department of Agriculture. Training of pesticide applicators is the function of the Cooperative Extension Service (FIFRA,Section 23 (c)).

13.10 The applicator certification program required by EPA is carried out in each state by the _____ _____.

Lead Agency

FIFRA was further amended in 1975, 1978, 1980, and 1981. These provisions clarified the intent of the law and will have a great influence on the way pesticides are registered and used. Many of the changes were designed to improve the registration process, which was slowed significantly by regulations resulting from the 1972 Act. The more important points are as follows:

1. Generic standards will be set for the active ingredients rather than for each formulated product. This change permits the EPA to make safety and health decisions

for the active ingredient in a pesticide, instead of treating each product on an individual basis. This will speed
registration, because there are only about 960 active
ingredients in the 25,000 formulations currently on the
market.
    2. Reregistration of all older products is required.
Under reregistration, all compounds are currently being
reexamined to make certain the supporting data for a registered pesticide satisfies today's requirements for registration, in light of new knowledge concerning human health
and environmental safety. Many pesticides were registered
under the old data requirements (prior to August 1975), and
registrants must submit new information to carry the product through the reregistration process.
    3. Pesticides can now be given *conditional registra-
tion*. EPA may grant a conditional registration for a pesticide even though certain supporting data have not been
completed. That information will still be required, but it
may be deferred to a later date. Conditional registration
may be granted by EPA if:
    a. The uses are identical or similar to those which
       exist on labels for already registered products
       with the same active ingredient;
    b. New uses are being added, providing a notice of
       Special Review (formerly referred to as RPAR) has
       not been issued on the product, or in the case of
       food or feed use, there is no other available or
       effective alternative;
    c. New pesticides have had additional data requirements
       imposed since the date of the original submission.
    4. Efficacy data may be waived. The EPA has the
option of setting aside requirements for proving the efficacy of a pesticide before registration. This leaves the
manufacturer to decide whether a pesticide is effective
enough to market, and final proof will depend on product
performance.
    5. The use of data from one registrant can be used by
other manufacturers or formulators if paid for. All data
provided from 1970 on can be used for a 15-year period by
other registrants, if they offer to pay "reasonable com-
pensation" for this use. In the future, registrants will
have 10 years of exclusive use of data submitted for a new
pesticide active ingredient. During that time, other
applicants may request and be granted permission to use the
information.
    6. Trade secrets will be protected. EPA may reveal
data on most pesticide effects (including human, animal,
and plant hazard evaluation), efficacy, and environmental
chemistry. Four categories of data are generally to be
kept confidential:
    a. Manufacturing and quality control processes;
    b. Methods of testing, detecting, or measuring
       deliberately added inert ingredients;
    c. Identity or quality of deliberately added inerts;

   d. Production, distribution, sale, and inventories of
      pesticides.
   7. The state now has primary enforcement responsibility, referred to as "state primacy." Under the 1978 law,
the primary authority for use enforcement under federal
law will be assigned to the states. Any suspected misuse
of pesticides will be investigated by and acted upon by the
state regulatory boards. Before the enforcement authority
can be legally transferred by EPA, the state must indicate
that their regulatory methods will meet or exceed the
federal requirements. If a state does not take appropriate
action within 30 days of alleged misuse, EPA can act. In
addition, enforcement authority can be taken away from any
state which consistently fails to take proper action.
   8. States can register pesticides for Special Local
Needs (SLN) (Section 24c). The state has authority to
register materials for unusual situations. Registrations
can be made for new products, using already registered
active ingredients. Existing product labels can be amended
for new uses, including chemicals which have been subject
to cancellation or suspension in the past. Only those
specific uses which are cancelled or suspended may not be
registered by the state for SLN's.
   9. Uses inconsistent with the labeling are defined.
The phrase, "to use any registered pesticide in a manner
inconsistent with its labeling" is defined.
   Users and applicators may now:
   a. Use a pesticide at less than labeled dosage, pro-
      viding the total amount applied does not exceed
      that currently allowed on the labeling;
   b. Use a pesticide for control of a target pest not
      named on the label, providing the site or host is
      indicated;
   c. Apply the pesticide using any method not specific-
      ally prohibited on the label;
   d. Mix one or more pesticides with other pesticides
      or fertilizers, provided the current labeling does
      not actually prohibit this practice.
   Remember! Other than these allowances, there must be
strict adherence to the label!

13.11 In the misuse of pesticides who or what has primary
      enforcement responsibility?
      -
      the state through its regulatory board

13.12 In what ways can a pesticide be used in a way that
      is inconsistent with its label?
      -
      less than labeled dosage; for pest not on label;
      application method not prohibited; mix with other
      pesticides not prohibited.

The Special Review process (formerly Rebuttable Presumption Against Registration (RPAR)) by the EPA is designed to ensure a full gathering of scientific information on pesticide safety and a thorough assessment of risks and benefits of pesticide products. This elaborate process allows EPA to study chemicals in depth before determining whether prolonged, courtroom-type hearings are necessary to cancel registrations or place restrictions on the uses of pesticides suspected or known to possess one or more of the risk criteria or "triggers" for Special Review. These risk criteria are concerned with the following:

1. Acute toxicity
2. Chronic toxicity
   a. Oncogenic
   b. Mutagenic
3. Other chronic effects
   a. Reproductive
      (1) Fetotoxicity
      (2) Teratogenicity
   b. Spermatogenicity
   c. Testicular effects
4. Significant reduction in wildlife, reduction in endangered species, and reduction in non-target species
5. Lack of emergency treatment or antidote

13.13 Special Review is designed to provide a thorough assessment of _____ and _____ of pesticides.

- -

risks and benefits

Most risk rebuttals are normally conducted by the pesticide's registrant, however, rebuttals may also be submitted by anyone (e.g., the U. S. Department of Agriculture (USDA), individual states, grower or commodity groups, and private parties.) In fact, the EPA may contest its own Reviews when appropriate. Thus, in the Special Review process, risks may be challenged by any interested party.

Benefits assessment and determination of exposure under use conditions are determined as a standard policy by the National Agricultural Pesticide Impact Assessment Program's (NAPIAP) assessment teams. The NAPIAP rebuttal, which involves every state, is at least as important as the EPA's Special Review process, for it provides a way for the people to be heard in the regulatory process. The assessment team is also charged with identifying short-term researchable data gaps.

13.14 How do farmers and other interested persons express
      their concerns and needs for pesticides in the
      Special Review process?
- -
      through the NAPIAP assessment teams

A pesticide tolerance is the maximum amount of a pesticide residue that can legally be present on a food or feed
at harvest or slaughter. The tolerance is expressed in
parts of the pesticide per million parts of the food or feed
by weight (ppm), and usually applies to the raw agricultural
commodity. Pesticide tolerances are set by the EPA and en-
forced by The Food and Drug Administration of The Department
of Health and Human Services or, in the case of meat, poultry and eggs, by USDA agencies.

The tolerance on each food is set sufficiently low
that daily .onsumption of the particular food or of all
foods treated with the particular pesticide will not result
in an exposure that exceeds the Acceptable Daily Intake
(ADI) for the pesticide. The tolerance is set still lower
if the effective use of the pesticide results in lower
residues.

The *Acceptable Daily Intake* (ADI) is defined as the
daily exposure level of a residue which during the entire
lifetime of man, appears to be without appreciable risk on
the basis of all facts known at the time. The ADI is
usually set one hundred times lower (1/100) than the No
Observable Effect Level (NOEL). A much greater safety
factor is required if there is evidence that the pesticide
causes cancer in test animals. Although the Delaney Amendment to the Federal Food, Drug and Cosmetic Act prevents
the addition of an animal carcinogen to foods it does not
apply to pesticide residues that occur inadvertently in
the production of the crop.

The *No Observable Effect Level* (NOEL) is the dosage
of the pesticide that results in no distinguishable harm
to experimental animals in chronic toxicity studies that
include the minute examination of all body organs for ab-
normality.

It is extremely unlikely that you will ever be exposed
to anything near the ADI in your food. Numerous, detailed,
continuing, nationwide studies show the actual pesticide
residues in food to be far below the ADI and the established tolerances.

The pesticide residues on crops at the time of harvest
are usually less than the tolerances. Residues decrease
during storage and transit. They are reduced further by
operations such as washing, peeling and cooking when the
food is prepared for eating. Every acre of a food crop
will not have been treated with the same pesticide, if any.

And you will not eat every food for which there is a tolerance for a particular pesticide every day. Further, the ADI refers to lifetime exposure. A minor excess of the ADI for a short period should be inconsequential.

13.15 In your own words, give definitions for these terms:
      Tolerance _____
      ADI   _____
      NOEL   _____

- -

      Tolerance - the amount of pesticide residue permitted
                  in food at harvest or slaughter;
      ADI   - the safe amount of pesticide that can be
                  consumed daily by humans;
      NOEL   - the highest daily dose of a pesticide
                  that does not cause harm in experimental
                  animals.

     These are only the most important aspects of FEPCA with which you, the interested novice, need be acquainted. Beyond the federal laws providing rather strict control over the use of pesticides, each state usually has several similar laws controlling the application, sale and use of pesticides. These may or may not involve the licensing of aerial and ground applicators as one group, and the structural applicator or pest control operator as another.

UNIT 14
SAFE HANDLING AND USE OF PESTICIDES

Pesticide chemicals are safe to use, provided common-sense safety is practiced and provided they are used according to the directions and precautions printed on the label; this includes keeping them away from children and illiterate or mentally incompetent persons.

14-A THE PESTICIDE LABEL

The single most important tool to the layman in the safe use of pesticides is the label on the container. The Federal Environmental Pesticide Control Act (FEPCA), which is discussed in Pesticides And The Law, contains three very important points concerning the pesticide label that I feel should be further emphasized. They pertain to reading the label, understanding the label directions, and following these instructions carefully.

Two of the first provisions of FEPCA are that (1) the use of any pesticide inconsistent with the label is pro-hibited, (except those listed on page 207), and (2) deliber-ate violations by growers, applicators, or dealers can result in heavy fines or imprisonment or both. The third provision is found in the general standards for certifica-tion of commercial applicators that in essence licenses them to use restricted-use pesticides, namely in the area of label and labeling comprehension. For certification, applicators are to be tested on (1) the general format and terminology of pesticide labels and labeling; (2) the understanding of instructions, warning, terms, symbols, and other information commonly appearing on pesticide labels; (3) classification of the product as restricted use; and (4) the necessity for use consistent with the label (with certain exceptions, page 207).

14.1 What is the most important tool for the layman in handling pesticides safely?

- -

the label

PANEL   H
ENVIRONMENTAL PROTECTION AGENCY FORMAT
FOR RESTRICTED-USE PESTICIDE LABEL
(Products bearing this kind of label
are not available ·to the layperson)

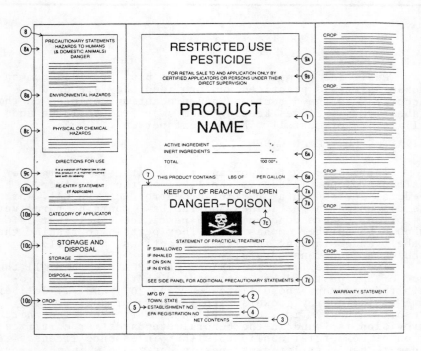

In Panel H is shown the format label for restricted-use pesticides as required by EPA.   This label is keyed as follows:

1. Product name
2. Company name and address
3. Net contents
4. EPA pesticide registration number
5. EPA formulator manufacturer establishment number
6A. Ingredients statement
6B. Pounds/gallon statement (if liquid)
7. Front-panel precautionary statements
7A. Child hazard warning, "Keep out of Reach of Children"
7B. Signal word - DANGER, WARNING, or CAUTION
7C. Skull and crossbones and word *POISON* in red
7D. Statement of practical treatment
7E. Referral statement
8. Side- or back-panel precautionary statements
8A. Hazards to humans and domestic animals

8B. Environmental hazards
8C. Physical or chemical hazards
9A. "Restricted Use Pesticide" block
9B. Statement of pesticide classification
9C. Misuse statement
10A. Reentry statement
10B. Category of applicator
10C. "Storage and Disposal" block
10D. Directions for use

Below are several recommended practices for the safe handling of pesticides. These may be useful to the reader in the actual use of pesticides around the home or on the farm, in preparation for talks on pesticide safety, safe working conditions, or just for reference.

14-B BASICS FOR THE EMPLOYER

1. Inform all employees of the hazards associated with pesticide use.
2. Have prearranged medical services available.
3. Post the telephone number and address of the nearest Poison Control Center at every telephone in the work situation.
4. Provide protective clothing and equipment; insist that it be worn when mixing, transferring or applying pesticides.

14.2 What is the name of the agency that can provide antidote information for poisoning incidents?

Poison Control Center

14-C SELECTION OF PESTICIDES

1. Identify the pest to be controlled, and if in doubt, consult your County Extension Agent or other authority.
2. Select the pesticide recommended by competent authority and make sure the label lists the crop or site to be treated.
3. Make certain that the label on container is intact and up-to-date; it should include directions and precautions.
4. Purchase only the quantity needed for the current season.

14.3 Whom should you consult regarding a particular pest and the materials recommended for its control?

the nearest County Extension Agent

14-D TRANSPORTING PESTICIDES

   1. Hazardous materials should be conspicuously marked
      on the container.
   2. Open-type vehicles are preferred to closed vehicles
      for transportation of volatile pesticides and those
      that would give off poisonous or noxious fumes if
      their containers were accidentally unsealed.
   3. All containers should be tightly closed. Don't
      haul ruptured bags, cans or drums.
   4. Always be prepared for an ACCIDENT.

14-E STORING PESTICIDES

   1. Lock all pesticide rooms, cabinets, or sheds.
   2. Don't store pesticides where food, feed, seed, or
      water can become contaminated.
   3. Store in a dry, well-ventilated place, away from
      sunlight, and at temperatures above freezing.
   4. Mark all entrances to storage, "PESTICIDES STORED
      HERE -- KEEP OUT" and if needed, "PELIGRO--ALMACEN
      DE VENEMOS--SE PROHIBE LA ENTRADA".
   5. Keep pesticides only in original containers, closed
      tightly and labeled.
   6. Examine pesticide containers frequently for leaks
      and tears. Dispose of leaking and torn containers,
      and clean up spilled or leaked material immediately.
   7. Where possible, a sink for washing should be
      located in or near storage.
   8. Keep an inventory and eliminate all outdated
      materials. Date containers when purchased.
   9. Take precautions for potential fire hazards.

14-F HANDLING AND MIXING PESTICIDES

   1. Before mixing, read carefully the label directions
      and current official state recommendations of your
      state's Cooperative Extension Service.
   2. Wear appropriate protective clothing and equipment,
      as specified on the label.
   3. Handle pesticides in a well-ventilated area. Avoid
      dusts and splashing when opening containers or
      pouring into the spray apparatus. Don's use or mix
      pesticides on windy days. Use closed transfer sys-
      tems, when available, for restricted use materials.
   4. Measure the quantity of pesticide required accu-
      rately, using the proper equipment.
   5. Don't mix pesticides in areas where there is a
      chance that spills or overflows could get into any
      water supply.
   6. Clean up spills immediately. Wash pesticides off
      skin promptly with plenty of soap and water.
      Change clothes immediately if they become contamin-
      ated.

14-G <u>APPLYING</u> <u>PESTICIDES</u>

1. Wear the appropriate protective clothing and equipment, as required for toxic materials.
2. Make certain equipment is calibrated correctly and
   is in satisfactory working condition.
3. Apply only at the recommended rate. To minimize
   drift, apply on a calm day and don't work through
   clouds or drift of unsettled dusts or sprays. Don't
   contaminate livestock, feed, food, or water supplies.
4. Avoid damage to beneficial and pollinating insects
   by not spraying during periods when such insects
   are actively working in the spray area. Notify
   neighboring beekeepers, at least 24 hours before
   application so precautionary measures can be taken.
   Honeybees are usually inactive at dawn and dusk
   which are good times for applications.
5. Keep pesticides out of mouth, eyes, and nose. Don't
   eat or smoke when using pesticides. Don't use mouth
   to blow out clogged lines or nozzles.
6. Observe precisely the *harvest intervals* specified
   between pesticide application and harvest and
   reentry intervals. Keep people and animals out of
   treated area as indicated on label.
7. Clean all equipment used in mixing and applying
   pesticides according to recommendations. Don't use
   herbicide-application equipment for applying insecticides.
8. After handling pesticides, wash the sprayer, protective equipment and hands thoroughly.
9. Keep complete and accurate records of the use of
   pesticides.
10. If symptoms of poisoning occur during or shortly
    after the use or exposure to a pesticide, call the
    physician or take the patient to the hospital im-
    mediately. Take the pesticide label with you.

14.4 Why keep records of pesticide use?

----

     for later reference, if neeced

14-H <u>DISPOSING</u> <u>EMPTY</u> <u>CONTAINERS</u> <u>AND</u> <u>UNUSED</u> <u>PESTICIDES</u>

Discarding empty containers which held pesticides is
a matter of concern to agricultural applicators, structural
pest control operators, and growers. Each state has developed its own specific pesticide container disposal regulations. The state lead agency for pesticides, as designated
by EPA, should be consulted for EPA-approved toxic waste
disposal sites and landfills. When this information is not
readily available, the following procedures are appropriate

and should be used.

   Pesticide containers which have been triple-rinsed and
punctured (making them unfit for further use) may be disposed of as non-hazardous wastes, in the normal manner.
They are not subject to any special regulations, other than
those imposed by the local sanitation agency or trash disposal firm. Large amounts should be taken to a sanitary
landfill, while smaller quantities may be placed in regular
trash containers.

   It is very important to remember, however, that the
water used   to rinse the containers must either be used
again to dilute or mix with pesticides, or disposed of as
a hazardous waste. Do not dump it!

   Glass containers which are triple-rinsed and crushed
may be disposed of in a sanitary landfill. Less than 50
pounds of empty paper containers may be burned in an open
area where local regulations permit.

14-I <u>INSTRUCTIONS</u> <u>FOR</u> <u>HOMEOWNERS</u>

   Because empty containers are never completely empty,
do not reuse them for any purpose. Instead, break glass
containers, triple-rinse metal containers with water, punch
holes in top and bottom and leave in your trash barrels for
removal to the municipal landfill. Empty paper bags and
cardboard boxes should be torn or smashed to make unusable,
placed in a larger paper bag, rolled, and relegated to the
trash barrel. Do not leave anything tempting in your trash.

14.5 Where can information be obtained regarding toxic
     waste and pesticide disposal sites and landfills?

----

     from the state lead agency for pesticides designated
     by EPA

14-J <u>UNWANTED</u> <u>PESTICIDES</u> <u>AND</u> <u>HAZARDOUS</u> <u>CONTAINERS</u>

   The most practical method for disposing of an unwanted
pesticide is to use it according to label directions. Next
best is to offer it to a responsible grower or neighbor in
need of the materials. If this is not practical, *homeowners*
may dispose of small quantities of pesticides by leaving
them in their original containers, wrapping in several
layers of newspapers and placing them in the trash. Do not
bury them; do not take to an incinerator and do not incin-
erate them yourself.

   Sacks which held inorganic pesticides or organic mercury, lead, cadmium, or arsenic compounds may not be burned,
as described above, but must be delivered to an EPA- or
state-approved Class I Treatment, Storage or Disposal (TSD)
Facility. In addition, empty unrinsed containers may not

be stored more than 90 days. This also applies to contam-
inated or useless pesticides, and disposal must be handled
as a hazardous waste, according to toxicity. They are to
be disposed of in an EPA- or state-approved Class I TSD
Facility.

14.6 How do you dispose of old pesticides that are no
     longer registered for use?

----

        deliver to a state-approved Class I Treatment,
        Storage or Disposal Facility

14-K SHELF LIFE

     Shelf life is the length of time a pesticide can be
stored and still give the expected control. Pesticides are
manufactured, formulated, and packaged to exacting standards. Almost without exception, pesticides are formulated
to provide a 2-year shelf life under normal storage conditions. However, they can deteriorate in storage, especially under conditions of high temperature and humidity. Some
pesticides lose active ingredients due to chemical decompo-
sition or volatilization. Dry formulations become caked
and compacted and the emulsifier in emulsifiable concentrates may have been inactivated. Some pesticides are converted into more toxic, flammable or explosive substances
as they decompose.
     Fewer problems occur with stored pesticides and the
products have a longer shelf life if the storage area is
cool, dry and out of direct sunlight. Protection from
temperature extremes is important because either condition
can shorten the shelf life of pesticides. At below-freez-
ing temperatures some liquid formulations separate into
various components and lose their effectiveness. High temperatures cause many pesticides to volatilize or break
down more rapidly. Extreme heat may also cause glass bottles to break or explode.
     Store granular pesticides on shelves if there is any
possibility of dampness on the floor. Separating volatile
herbicides and other pesticides is a wise precaution
against cross-contamination. Keep all corrosive chemicals
in proper containers to prevent leaks that might result in
serious damage. Even the simple step of tightly closing
lids and bungs on containers can help extend the shelf life
of pesticides.
     Over a period of time containers may develop leaks,
breaks, or tears, so check them often for such problems.
If a damaged container is found, transfer its contents to
a clearly labeled overpack container or to one that held
the same formulation previously. Don't tear open the tops
of new bags or boxes of pesticides. Keep a sharp knife

handy for this purpose, and clean it each time a container is opened. Partly empty paper containers should be sealed with tape or staples.

When you buy pesticides, date them and keep a current inventory of your supplies. Avoid stockpiling; buy what you need, but not to excess. This eliminates waste and the problem of what to do with old materials.

Sometimes it is necessary to carry pesticide stocks over from one year to the next. Check dates of purchase at the beginning of each season, and use the older materials first. To protect the label on a container and to keep it intact and legible, cover it with transparent tape or lacquer.

If given proper storage, pesticides may remain active for several years. However, most pesticides are not backed by the manufacturer if stored longer than two years; so plan every purchase of pesticides to be completely used within this two year period.

14.7 The shelf life intended for most pesticide formula-
     tions by their manufacturers is _____ _____.

- -

     two years

14-L PESTICIDE EMERGENCIES

Despite the most thorough precautions, accidents will occur. The following three sources of information are important in the event of any kind of serious pesticide problem.

The first and most important source of information is the CHEMTREC telephone number. From this toll-free long-distance number can be obtained emergency information on all pesticide accidents, pesticide-poisoning cases, pesticide spills, and pesticide-spill cleanup teams. This telephone service is available 24 hours a day. The toll-free number is:

CHEMTREC (800) 424-9300

The second source of information is only for human-poisoning cases: It is the nearest Poison Control Center. Look it up in the telephone directory under Poison Control Centers, or ask the telephone operator for assistance. Poison Control Centers are usually located in the larger hospitals of most cities and can provide emergency treatment information on all types of human poisoning, including pesticides. The telephone number of the nearest Poison Control Center should be kept as a ready reference by parents of small children or employers of persons who work with pesticides and other potentially hazardous materials.

Specific pesticide poisoning information can be obtained from your nearest Regional Poison Control Centers. There are presently 32 of these scattered across the U.S., with 24-hour professional services at each. Following is a list of these and their direct line phone numbers, some of which have 800 toll-free numbers.

REGIONAL POISON CONTROL CENTERS

ALABAMA, Tuscaloosa
800-462-0800 (Alabama only)
(205) 345-0600

ARIZONA, Tucson
800-362-0101 (Arizona only)
(602) 626-6016

CALIFORNIA, Sacramento
(916) 453-3414

CALIFORNIA, San Diego
(619) 294-6000

CALIFORNIA, San Francisco
800-233-3360
(415) 666-2845

CALIFORNIA, Los Angeles
(213) 484-5151

COLORADO, Denver
(303) 893-7774

FLORIDA, Tampa
800-282-3171
(813) 251-6995

GEORGIA, Atlanta
(404) 589-4400

ILLINOIS, Springfield
800-252-2022
(217) 753-3330

INDIANA, Indianapolis
800-382-9097
(317) 630-7351

IOWA, Iowa City
(319) 356-2922

KENTUCKY, Louisville
800-722-5725
(502) 562-7270

LOUISIANA, Shreveport
800-535-0525
(318) 425-1524

MARYLAND, Baltimore
(301) 528-7701

MICHIGAN, Detroit
(313) 494-5711

MICHIGAN, Grand Rapids
800-442-4571 (AC 616 only)
800-632-2727 (Michigan only)

MINNESOTA, Minneapolis
(612) 347-3141

MINNESOTA, St. Paul
800-222-1222
(612) 221-2113

MISSOURI, St. Louis
(314) 772-5200

NEBRASKA, Omaha
800-642-9999
(402) 390-5434

NEW JERSEY, Newark
800-962-1253 (New Jersey only)
(201) 926-8005

NEW MEXICO, Albuquerque
800-432-6866
(505) 843-2551

NEW YORK, Long Island
(516) 542-2323

NEW YORK, New York
(212) 340-4494

NORTH CAROLINA, Durham
800-672-1697 (North Carolina only)
(919) 684-8111

OHIO, Cincinnati
(513) 872-5111

OHIO, Columbus
(614) 461-2012
(614) 228-1323

PENNSYLVANIA, Pittsburgh
(412) 647-5600

UTAH, Salt Lake City
(801) 581-2151

WASHINGTON, D. C.
(202) 625-3333

WASHINGTON, Seattle
(206) 526-2121

14.8 The three sources of pesticide emergency information
     that can be had by calling a toll free number are:
     _____, _____,
     and _____.

     CHEMTREC, local Poison Control Centers, and Regional
     Poison Control Centers.

     In summary, any pesticide can be used safely if com-
mon sense safety is practiced and the label instructions
on the container are followed carefully.

UNIT 15
THE TOXICITY OF PESTICIDES

        Pesticides, by necessity, are poisons, but the toxic-
ity and hazards of different compounds vary greatly. There
is a great distinction between toxicity and hazard. *Toxicity* refers to the inherent toxicity of a compound - in
other words, to how toxic it is to animals under experimental conditions. *Hazard* refers to the risk or danger of
poisoning when a chemical is used or applied, sometimes
referred to as *use hazard*.
        The user of a pesticide is really concerned with the
use hazard and not the inherent toxicity of the material.
Hazard depends not only on toxicity but also on the chance
of exposure to toxic amounts of the pesticide.
        As far as the possible risks associated with the use
of pesticides are concerned, we can distinguish between two
types: First, *acute poisoning*, resulting from the handling
and application of toxic materials; and second, *chronic
poisoning* from long-term exposure to small quantities of
materials or from ingestion of them. The question of acute
toxicity is obviously of paramount interest to people
engaged in manufacturing and formulating pesticides and to
those responsible for their application. Supposed chronic
risks, however, are of much greater public interest because
of their potential effect on the consumer of agricultural
products.

15.1 What are the two types of risks associated with the
     use of pesticides?

- -

        acute poisoning   chronic poisoning

        Fatal human poisoning by pesticides is uncommon in the
United States and is due to accident, suicide, or on rare
occasion, crime. Fatalities represent only a small frac-
tion of all recorded cases of poisoning, as demonstrated by
these statistics for 1983 (Panel J). Note that 2.6 percent
of the deaths from poisoning were from pesticides. Of all
poison exposure victims, that is reported incidents, and
not just fatalities, 64% are children under 6 years of age.
The picture for pesticides is very little better, with

222   THE TOXICITY OF PESTICIDES

61.6% of the reported poison exposure cases being children
under 6 years of age.

   Regardless of your attitudes toward pesticides and
their exaggerated hazard, they have a very respectable
safety record, which is improving each year, mainly through
education, container labeling, and child-proof container
lids.

PANEL J
TOTAL ESTIMATED DEATHS FROM POISONING
(ACCIDENTAL, INTENTIONAL, ADVERSE REACTION)
IN THE U. S. DURING 1983[a]

Alcohols	109[b]
Automotive/aircraft/boat products	18
Bites and envenomations	18
Chemicals	91
Cleaning substances	36
Cosmetics/personal care products	9
Fumes/gases/vapors	45
Heavy metals	9
Hydrocarbons	18
Analgesics (aspirins, acetaminophen, etc.)	182
Antidepressants	164
Antimicrobials	18
Asthma therapies	36
Cardiovascular drugs	45
Cough and cold preparations	9
Electrolytes/minerals	27
Pesticides	27
Sedatives/hypnotics/antipsychotics	100
Stimulants and street drugs	64
Topical agents	9
Unknown drugs	9
Total	1,043

[a] Source: Veltri, J. A. and T. L. Litovitz. 1984. 1983
   Annual Report of the American Association of
   Poison Control Centers National Data Collection
   System. American Journal of Emergency Medicine.
   2(5):419-443.

[b] The authors stated, "Noting the 234.0 million total
   United States population during this period, the data
   presented (25.8 million) represent an estimated 11% of
   the human poison exposures reported to poison control
   centers in the United States each year." I have taken
   the liberty of extrapolating their data from the 11%
   represented to the full population. The numbers shown
   above represent an estimate of the total fatal poisoning
   cases based on 11% of the total population.

15.2 What group of chemicals causes most of the fatal
ingestion poisonings? (Take your answer from Panel J).
----
analgesics

15-A HOW PESTICIDE TOXICITY IS DETERMINED

An understanding of the basic principles of toxicity
and the differences between toxicity and hazard is essential. As you already know, some pesticides are much more
toxic than others, and severe illness may result when only
a small amount of a certain chemical has been ingested,
while with other compounds no serious effects would result
even after ingesting large quantities. Some of the factors
that influence this are related to (1) the toxicity of the
chemical, (2) the dose of the chemical, especially concentration, (3) length of exposure, and (4) the route of entry
or absorption by the body.
Present requirements for testing the product to evaluate any hazards that might be related to it include the
following general categories of tests on animals or in the
environment:

Acute oral, dermal, and inhalation
Eye and skin irritation
Neurotoxicity (effects on the nervous system)
Reproduction
Teratogenicity (birth defects)
Mutagenicity (genetic effects)
Dermal sensitization
Subchronic feeding (90 days - two species)
Chronic feeding (2 years - carcinogenicity/cancer-
two rodent species)
Chronic feeding (1 year-one nonrodent species)
Wildlife effects
Environmental fate

The law places the burden of proving the safety of
pesticide usage on the manufacturer. For this reason, over
the past 30 years the hazard evaluation studies have generally been conducted by scientific laboratories maintained
by the manufacturers or through outside contract laboratories.
The U. S. EPA and other regulatory agencies world-wide
require a battery of toxicity studies before approval of
the use of a pesticide in the production of food crops.
Those key toxicity categories are:

Acute studies which evaluate toxicity from single,
short-duration exposures, include several tests which

address lethality by oral, dermal, or inhalation contact;
irritation by dermal or eye contact; and sensitization by
dermal contact.

Subchronic and chronic studies which evaluate toxicity
from multiple or continuous long-term exposure, include
ninety-day dog and rat subchronic studies and one-year dog,
eighteen-month mouse, and two-year rat chronic studies.

Reproductive effect studies which evaluate potential
impairment of reproductive function and fetal development,
include a two-generation rat reproduction study and two
teratology studies generally conducted with rats and rabbits.

Mutagenic studies which evaluate potential structural
or functional impairment of genetic material, usually include a battery of four or more assays specific for identifying DNA and chromosome damage.

These tests are conducted on test animals that are
relatively easy to work with and whose physiology, in some
instances, is like that of humans; for example, white mice,
white rats, white rabbits, guinea pigs, and beagle dogs.
For instance, intravenous tests are determined usually on
mice and rats, whereas dermal tests are conducted on shaved
rabbits and guinea pigs. Acute oral toxicity determina-
tions are most commonly made with rats and dogs, the test
substance being introduced directly into the stomach by
tube. Chronic studies are conducted on the same two species
for extended periods, and the compound is usually incorporated in the animal's daily ration. Inhalation studies
may involve any of the test animals, but rats, guinea pigs,
and rabbits are most commonly used.
These procedures are necessary to determine the overall toxic properties of the compound to various animals.
From this information, toxicity to man can generally be
extrapolated, and eventually some micro-level portion of
the pesticide may be permitted in his food as a residue,
which is expressed in ppm.
Toxicologists use rather simple animal toxicity tests
to rank pesticides according to their toxicity. Long before pesticides are registered with the EPA and eventually
released for public use, the manufacturer must declare the
toxicity of their pesticide to the white rat under laboratory conditions. This toxicity is defined by the LD_{50},
expressed as milligrams (mg) of toxicant per kilogram (kg)
of body weight, the dose that kills 50 percent of the test
animals to which it is administered under experimental
conditions.
The LD_{50} is measured in terms of oral (fed to, or
placed directly in the stomachs of rats), dermal (applied
to the skin of rats and rabbits), and respiratory toxicity
(inhaled). The size of the dose is the most important

single item in determining the safety of a given chemical.

15.3 Why are rats, rabbits, guinea pigs and dogs used to
     estimate toxicity of chemicals to man?
- -
     easy to work with and their physiology is like that
     of humans

15.4 Name 5 types of toxicity data collected before a
     pesticide is registered for use, using the list on
     page 223.
- -
     there are 10 to choose from

15.5 The term LD_{50} means: _____.
- -
     lethal dose to 50% of the test animals

15.6 In what units or measurements are LD_{50}s expressed:
- -
     mg/kg (milligrams of compound per kilogram of animal)

15-B ESTIMATING TOXICITY TO HUMANS
     Besides toxicity, the other important variables are
the dose, length of exposure, and route of absorption. The
amount of pesticide required to kill a human being can be
correlated with the LD_{50} of the material to rats in the
laboratory. In Panel K, for example, the acute oral LD_{50}
expressed as mg/kg dose of the technical material, is trans-
lated into the amount needed to kill a 170-pound human.
Dermal LD_{50}s are included for a better understanding of the
expressed animal toxicity to human toxicity.
     Generally speaking, oral ingestions are more toxic
than respiratory inhalations, which are more toxic than
dermal absorption. However, workers' exposure is usually
dermal, which explains why most illnesses are reported in
workers who have skin contact with pesticides. Consequently, dermal toxicity information, or the dermal LD_{50},
is of more value in determining the hazard of a pesticide
to workers than the oral LD_{50}.

15.7 Of the two types of available pesticide toxicity data,
     which is more meaningful to the user, and why?
- -
     dermal: The user is most likely to be exposed dermally.

PANEL K
COMBINED TABULATION OF TOXICITY CLASSES

Routes of Absorption

Toxicity Rating	Commonly Used Term	LD50 Single ORAL Dose Rats mg/kg	LD50 Single DERMAL Dose Rabbits mg/kg	Probable Lethal Oral Dose for Man
6	Supertoxic	< 5	< 20	A taste, a grain
5	Extremely toxic	5-50 mg	20-200	A pinch, 1 teaspoon
4	Very toxic	50-500	200-1,000	1 teaspoonful - 2 tablespoons
3	Moderately toxic	500-5,000	1,000-2,000	1 ounce - 1 pint
2	Slightly toxic	5,000-15,000	2,000-20,000	1 pint - 1 quart
1	Practically non-toxic	> 15,000	> 20,000	> 1 quart

< means LD50 is less than the figure shown.
> means LD50 is higher than the figure shown.

Source: Toxicity ratings modified from M. N. Gleason, R. E. Gosselin, and H. C. Hodge. 1976. Clinical Toxicology of Commercial Products. 4th ed. Williams and Wilkins Company. Baltimore, Md. p. 6.

PANEL L
EPA LABELING TOXICITY CATEGORIES BY HAZARD INDICATOR

Hazard Indicators	Toxicity Categories			
	I	II	III	IV
Oral LD_{50}	Up to and including 50 mg/kg	From 50 thru 500 mg/kg	From 500 thru 5000 mg/kg	Greater than 5000 mg/kg
Inhalation LD_{50}	Up to and including 0.2 mg/liter	From 0.2 thru 2 mg/liter	From 2 thru 20 mg/liter	Greater than 20 mg/liter
Dermal LD_{50}	Up to and including 200 mg/kg	From 200 thru 2000	From 2000 thru 20,000	Greater than 20,000
Eye effects	Corrosive; corneal opacity not reversible within 7 days	Corneal opacity reversible within 7 days; irritation persisting for 7 days	No corneal opacity; irritation reversible within 7 days	No irritation
Skin effects	Corrosive	Severe irritation at 72 hours	Moderate irritation at 72 hours	Mild or slight irritation at 72 hours

15.8 If the LD_{50} for compound X is 500 mg/kg, how much
would be required to achieve an LD_{50} on a man
weighing 70 kg?

- -

70 X 500   (35,000 mg)

15-C TOXICITY CATEGORIES AND SIGNAL WORDS ON PESTICIDE LABELS

The pesticide label must contain "signal words" in
bold print: *DANGER - POISON: WARNING;* and *CAUTION*. These
are significant words, since they represent an EPA category of toxicity, and thus give an indication of their
potential hazard (Panel L). There are four EPA categories
for pesticides:

Category I.   The signal words *DANGER - POISON* and the
skull and crossbones symbol are required
on the labels for all highly toxic compounds. These pesticides all fall within
the acute oral LD_{50} range of 0 to 50 mg/
kg.

Category II.   The word *WARNING* is required on the
labels for all moderately toxic compounds.
They all fall within the acute oral LD_{50}
range of 50 to 500 mg/kg.

Category III. The word *CAUTION* is required on labels
for slightly toxic pesticides that fall
within the acute oral LD_{50} range of 500
to 5,000 mg/kg.

Category IV.   The word *CAUTION* is required on labels
for compounds having acute oral LD_{50}s
greater than 5,000 mg/kg. However,
unqualified claims for safety are not
acceptable on any label, and all labels
must bear the statement, "Keep Out Of
Reach of Children."

In Panel M are listed several examples of insecticides,
herbicides, and fungicides found in the first three EPA
pesticide categories.

PANEL M

EXAMPLES OF INSECTICIDE TOXICITY CLASSES

Label Classification	Oral LD$_{50}$	mg/kg	Dermal LD$_{50}$	mg/kg
Extremely toxic	aldicarb, Temik[R]	0.65-0.79	parathion	7-21
	fensulfothion, Dasanit[R]	4.7-10.5	mevinphos, Phosdrin[R]	4.2-4.7
	monocrotophos, Azodrin[R]	21	disulfoton, Di-syston[R]	2.6-8.6
	phorate, Thimet[R]	1.1-2.3	demeton, Systox[R]	2.5-6.0
Moderately toxic	propoxur, Baygon[R]	95-104	methyl parathion	67
	chlorpyrifos, Dursban[R]	135-163	dioxathion, Delnav[R]	63-235
	diazinon	300-850	azinphosmethyl, Guthion[R]	220
Slightly toxic	malathion	1,000-1,375	toxaphene	780-1,075
	carbaryl, Sevin[R]	500-850	fenvalerate, Pydrin[R]	>5,000
	permethrin, Ambush[R], Pounce[R]	450->4,000	dicofol, Kelthane[R]	1,000-1,230
	temophos, Abate[R]	2,030	malathion	>4,444
	stirofos, Gardona[R]	4,000	carbaryl, Sevin[R]	>4,000

> means LD$_{50}$ is higher than figure shown

230

PANEL M (Continued)
EXAMPLES OF HERBICIDE TOXICITY CLASSES

Label Classification	Oral LD$_{50}$	mg/kg	Dermal LD$_{50}$	mg/kg
Highly toxic	DNOC sodium arsenite	25-40 10-50	None	
Moderately toxic	2,4-D paraquat	375 157	paraquat acifluorfen, Blazer^R	236-480 450
Slightly toxic	MSMA monuron simizine, Princep^R pendimethalin, Prowl^R	700-1,800 2,300-3,700 5,000 1,250	endothall dichlobenil 2,4-D acid MCPA	750 500 1,500 >1,000

> means LD$_{50}$ is higher than figure shown

PANEL M (Continued)
EXAMPLES OF FUNGICIDE TOXICITY CLASSES

Label Classification	Oral LD$_{50}$	mg/kg	Dermal LD$_{50}$	mg/kg
Highly toxic	cycloheximide, Actidione[R]	1.8-2.5	None	
	fentin chloride, Tinmate[R]	18		
Moderately toxic	binapacryl, Morocide[R]	136-225	triphenyltin acetate, Brestan[R]	500
	triphenyltin hydroxide	108		
Slightly toxic	thiram	780	binapacryl, Morocide[R]	720-810
	anilazine, Dyrene[R]	2,710	dinoseb	500
	ethazol, Koban[R]	1,077	maneb	>1,000
	dimethirimol, Milcurb[R]	2,350	zineb	>1,000
	dicloran, Botran[R]	5,000	triphenyltin hydroxide, Du-Ter[R]	5,000

> means LD$_{50}$ is higher than figure shown

The relative acute toxic hazards to applicators of
many of the most commonly used pesticides are shown in
Panel N. This is a valuable reference. Examples of
insecticides, herbicides and fungicides are included.
Notice, however, that these toxicity categories are not
related to specific categories just listed for label
requirements. Rather, they are relative toxicities, listed
from most to least dangerous.

With regard to the classifications of pesticides,
their general toxicities, in decreasing order, would be
insecticides > defoliants > desiccants > herbicides > fungicides. Within the most toxic class, the insecticides, the
categories would fall in the following general order of
their dermal hazards to humans: organophosphates > carbamates > cyclodienes > DDT-relatives > botanicals > activators
or synergists > inorganics. There are usually exceptions in
each category listed.

The formulation of pesticides, because of their varying kinds of diluents, would also have varying degrees of
hazard to humans. Again, we must generalize: liquid pesticide > emulsifiable concentrate > oil solution > water emulsion > water solution > wettable powder/flowable (in suspension) > dust > granular.

The petroleum solvents used in formulating pesticides
are also toxic. Currently these are diesel fuel, deodor-
ized kerosene, methanol, petroleum distallates, xylene, and
toluene. Of these, only the xylene and toluene are aromatics, and they offer by far the greater dermal hazard.

15.9 Which class of solvents offers the greatest dermal
     hazard to man?

- -

     aromatics

15.10 Which group or class of pesticides would generally
      be classed most toxic to man?

- -

      insecticides

15.11 Now, of this group, which group would be most toxic
      to man?

- -

      organophosphates

15.12 Of the three types of toxicity data, oral, dermal,
      and intravenous, which gives the more realistic
      estimate of worker hazard?

      dermal

PANEL N
ESTIMATED RELATIVE ACUTE TOXIC HAZARDS
OF PESTICIDES TO APPLICATORS[a/]

Most Dangerous	Dangerous
aldicarb, Temik[R] (I)[b/] demeton, Systox[R] (I) disulfoton, Di-Syston[R] (I) endrin (I) fensulfothion, Dasanit[R] (I) fonofos, Dyfonate[R] (I) mevinphos, Phosdrin[R] (I) parathion (I) phorate, Thimet[R] (I) terbufos, Counter[R] (I)	carbophenothion, Trithion[R] (I) cycloheximide (F) dialifor, Torak[R] (I) dichlorvos, Vapona[R] (I) dicrotophos, Bidrin[R] (I) dieldrin (I) dinitrophenol (F,I) dioxathion, Delnav[R] (I) dinoseb, DNBP (H,D,I) DNOC (I,F,H,D) EPN (I) ethion, Nialate[R] (I) fenaminosulf, Lesan[R] (F) isofenfos, Amaze[R] (I) methamidophos, Monitor[R] (I) methyl parathion (I) monocrotophos, Azodrin[R] (I) nicotine sulfate (I) paraquat (H,D) pentachlorophenol (D,F) phosphamidon, Dimecron[R] (I) sodium arsenite (H,I)

Less Dangerous	Least Dangerous
acifluorfen, Blazer[R] (H) aminocarb, Matacil[R] (I) azinphosmethyl, Guthion[R] (I) bendiocarb, Ficam[R] (I) binapacryl, Morocide[R] (F, I) bufencarb, Bux[R] (I) bupirimate, Nimrod[R] (F) butrizol, Indar[R] (F) chlordane (I) chlordimeform, Fundal[R], Galecron[R](I) coumaphos, Co-Ral[R] (I) crotoxyphos, Ciodrin[R] (I) diazinon (I) dimethoate, Cygon[R] (I) endosulfan, Thiodan[R] (I) endothall (H) fenthion, Baytex[R] (I) lindane (I) metam-sodium, Vapam[R](fum.,F,H) methidathion, Supracide[R] (I) methomyl, Lannate[R], Nudrin[R] (I)	alachlor, Lasso[R] (H) atrazine (H) bensulide, Prefar[R] (H) captan (F) carbaryl (I) carbofuran, Furadan[R] (I) chlorpyrifos, Dursban[R], Lorsban[R] (I) copper, organic and inorganic (F) cyhextin, Plictran[R] (I) cypermethrin, Ammo[R], Cymbush[R] (I) 2,4-D (H) DEF[R], merphos (D) dinocap (F) diquat (H) fenvalerate, Pydrin[R] (I) flucythrinate, Pay-Off[R] (I) fluvalinate, Mavrik[R], Spur[R] (I) folpet, Phaltan (F) formetanate, Carzol[R] (I) malathion (I) maneb (F)

234

PANEL N (Continued)
ESTIMATED RELATIVE ACUTE TOXIC HAZARDS
OF PESTICIDES TO APPLICATORS[a/]

Less Dangerous	Least Dangerous
naled, Dibrom[R] (I)	metalaxyl, Ridomil[R], Subdue[R] (F)
oxamyl, Vydate[R] (I)	oxythioquinox, Morestan[R] (F)
oxydemetonmethyl, Meta-Systox[R]-R (I)	PCNB, Terraclor[R] (F)
profenofos, Curacron[R] (I)	permethrin, Ambush[R], Pounce[R] (I)
propachlor, Bexton[R] (H)	phosalone, Zolone[R] (I)
prosulfalin, Sward[R] (H)	phosmet, Imidan[R] (I)
sulprophos, Bolstar[R] (I)	pyramdron, Amdro[R], Maxforce[R] (I)
tebuthiuron, Spike[R] (H)	simazine (H)
temephos, Abate[R] (I)	stirofos, Gardona[R] (I)
toxaphene (I)	tetradifon, Tedion[R] (I)
trichlorfon, Dipterex[R], Dylox[R] (I)	thiodicarb, Larvin[R] (I)
triphenyltin acetate, Brestan[R] (F)	thiram (F)
	thidiazuron, Dropp[R] (D)
	triadimefon, Bayleton[R] (F)
	trifluralin, Treflan[R] (H)
	zineb (F)
	ziram (F)

a/ The estimates of hazards in this table are based primarily on the observed acute dermal and to a less extent oral toxicity of these compounds to experimental animals. Where it is available, use experience has also been considered. It should be noted that the classification into toxicity groups is both approximate and relative. These toxicity categories are not related to specific categories spelled out for label requirements.

b/ The pesticide category to which the chemical belongs is designated as follows: D, defoliant; F, fungicide; H, herbicide; and I, insecticide/miticide.

EPA requires safety waiting intervals between application of certain insecticides and worker reentry into
treated fields, to prevent unnecessary dermal exposure.
Reentry intervals have been established only for insecticides because they pose the greatest health hazard to agricultural workers. Several states (for example, California)
have adopted waiting intervals longer than those required
by EPA. The established waiting intervals are:

48 hours
    Chlorpyrifos (Lorsban^R)
    Ethyl parathion
    Methyl parathion
    Demeton (Systox^R)
    Monocrotophos (Azodrin^R)
    Carbofenothion (Trithion^R)
    Oxydemetonmethyl (Metasystox-R^R)
    Dicrotophos (Bidrin^R)
    Endrin

24 hours
    Azinphosmethyl (Guthion^R)
    Phosalone (Zolone^R)
    EPN
    Ethion

In addition to reentry intervals established by the
EPA, several manufacturers have set reentry intervals on
certain of their insecticides:

48 hours
    Sulprofos (Bolstar^R)

24 hours
    Acephate (Orthene^R)
    Fenvalerate (Pydrin^R)
    Methomyl (Lannate^R and Nudrin^R)
    Permethrin (Ambush^R and Pounce^R)
    Phosphamidon

For all other insecticides, it is necessary only that
workers wait until sprays have dried or dusts have settled
before reentering treated fields. These worker safety
intervals are not to be confused with the familiar harvest
intervals - the minimum days from last treatment to harvest
- indicated on the insecticide label.
    If workers must enter fields earlier than the required
waiting intervals, they must wear protective clothing.
This consists of a long-sleeved shirt, long trousers or
coveralls, hat, shoes, and socks.
    These waiting intervals should not impose any undue
hardship on pest-management specialists and agricultural
pest control advisors, because application of any one of
these materials would preclude the necessity for field
inspection within the required waiting intervals.

15.13 Reentry intervals were established to prevent
      unnecessary _____ exposure of field workers
      to pesticides.
----
      dermal

15-E RESTRICTED-USE PESTICIDES

In the Unit on Safe Handling and Use of Pesticides, it
is stated that there are only two classes of pesticides
registered with the EPA, *restricted-use* and *unclassified*.
Unclassified pesticides may be purchased and applied by any
person. Restricted-use pesticides can be purchased and
applied only by certified applicators, persons having received special training and tested in the use, handling,
safety and application of pesticides. This training and
testing are administered in each state by an agency authorized by EPA to certify applicators, usually the same one
that licenses commercial pesticide applicators.

The criteria for restricted-use classification are
usually based on human hazard; however, others include
effects on aquatic organisms, effects of residues on birds,
hazard to nontarget organisms, and accident history.

The EPA prepared a list of 73 pesticides classified as
restricted-use as of September, 1985. An updated version
of this list is presented in Panel O. Note that not all
formulations of all 73 materials are restricted use.

Because this list will be changed from time to time it
can be considered incomplete. For the most recent additions
contact your County Extension Agricultural Agent or the
State Extension Pesticide Training Specialist.

15.14 Why are some formulations of a particular pesticide
      placed in the restricted-use category and some are
      not?
----
      not all formulations have the same potential hazard

PANEL O
RESTRICTED USE PESTICIDES AS OF SEPTEMBER 1985

Pesticide	Restricted Uses
acrolein, Aqualin^R (H)$^{a/}$	all
acrylonitrile, Acrylon^R (F)	all
aldicarb, Temik^R (I)	ornamental uses
allyl alcohol (H)	all
aluminum phosphide, Phostoxin^R (F)	all
amitraz, BAAM^R (I,M)	all
azinphosmethyl, Guthion^R (I)	concentrations greater than 13.5%
brodifacoum, Talon^R (R)	all
bromadiolone, Maki^R (R)	all
calcium cyanide, Cyanogas^R (F)	all
carbofuran, Furadan^R (I)	concentrations 40% and greater
chlordimeform, Fundal^R, Galecron^R(I)	all
chlorfenvinphos, Supona^R (I)	concentrations 21% and greater
chlorobenzilate (A)	use only on citrus
chlorophacinone, Rozol^R (R)	all
chloropicrin, Dowfume^R (F)	all
clonitralid, Bayluscide^R (M)	all
cyanazine, Bladex^R (H)	all
cycloheximide, Acti-Aid^R (B)	all
cypermethrin, Ammo^R and Cymbush^R (I)	all
DBCP (N)	use only on pineapple
demeton, Systox^R (I)	1% fertilizer & 2% granular
diallate, Avadex^R (H)	all
diclofop methyl, Hoelon^R (H)	all
dicrotophos, Bidrin^R (I)	all
diflubenzuron, Dimilin^R (I)	all
dioxathion, Delnav^R (I)	concentrations greater than 30%
disulfoton, Di-Syston^R (I)	formulations greater than: EC 65%, 95% oil solution, 10% granular

PANEL O (Continued)
RESTRICTED USE PESTICIDES AS OF SEPTEMBER 1985

Pesticide	Restricted Uses
dodemorph, Milban[R] (F)	all
endrin (I)	all
EPN (I)	all
ethoprop, Mocap[R] (I,N)	all
ethyl parathion (I)	all
fenamiphos, Nemacur[R] (I)	all
fensulfothion, Dasanit[R] (N,I)	all
fenvalerate, Pydrin[R] (I)	all
fluoroacetamide, 1081 (R)	all
flucythrinate, Pay-Off[R] (I)	all
fonofos, Dyfonate[R] (I)	all
heptachlor (I)	all
hydrocyanic acid, HCN (F)	all
isofenphos, Amaze[R] (I)	all
lindane (I)	all
magnesium phosphide (F)	all
methamidophos, Monitor[R] (I)	all
methidathion, Supracide[R] (I)	all
methiocarb, Mesurol[R] (I,M)	all
methomyl, Lannate[R], Nudrin[R] (I)	all formulations except 1% fly bait
methyl bromide (F)	containers greater than 1.5 lbs
methyl parathion (I)	all
mevinphos, Phosdrin[R] (I)	all
monocrotophos, Azodrin[R] (I)	all
nicotine (alkaloid) (I)	all
nitrofen, TOK[R] (H)	all
paraquat (H) (dichloride and bis methyl sulfate)	all formulations except: 0.44% pressurized spray; and in liquid fertilizer at 0.025%, 0.03% or 0.04% with atrazine

PANEL 0 (Continued)
RESTRICTED USE PESTICIDES AS OF SEPTEMBER 1985

Pesticide	Restricted Uses
permethrin, Ambush^R, Pounce^R, (I)	all
phorate, Thimet^R (I)	all
phosacetim, Gophacide^R (R)	all
phosphamidon, Dimecron^R (I)	all
picloram, Tordon^R (H)	all formulations except Tordon^R 101R
profenofos, Curacron^R (I)	all
pronamide, Kerb^R (H)	all
propetamphos, Safrotin^R (I)	all
sodium cyanide, Cymag^R (R)	all capsules & ball formulations
sodium fluoroacetate (R) Compound 1080	all
Starlicide^R (avicide)	all
strychnine (P)	all formulations & baits greater than 0.5%
sulfotep, Bladafum^R (I)	all spray & smoke generator uses
sulprofos, Bolstar^R (I)	all
tepp (I)	all
terbufos, Counter^R (I,N)	all
toxaphene (I)	all
zinc phosphide (R)	all

a/ The pesticide category to which the chemical belongs is designated as follows:

F, fumigant; H, herbicide; I, insecticide/acaricide; M, molluscicide; N, nematicide; P, predicide; R, rodenticide; B, bactericide; and A, acaricide.

15-F FIRST AID FOR POISONING

Poisoning symptoms may appear almost immediately after exposure or may be delayed for several hours, depending on the chemical, dose, length of exposure, and the individual. Symptoms may include, but are not restricted to, headache, giddiness, nervousness, blurred vision, cramps, diarrhea,

a feeling of general numbness, or abnormal size of eye
pupils. In some instances, there is excessive sweating,
tearing, or mouth secretions. Severe cases of poisoning
may be followed by nausea and vomiting, fluid in the lungs,
changes in heart rate, muscle weakness, breathing difficulty, confusion, convulsions, coma, or death. However,
pesticide poisoning may mimic brain hemorrhage, heat stroke,
heat exhaustion, hypoglycemia (low blood sugar), gastro-
enteritis (intestinal infection), pneumonia, asthma, or
other severe respiratory infections.

Regardless of how trivial the exposure may seem, if
poisoning is present or suspected, obtain medical advice
at once. If a physician is not immediately available by
phone, take the person directly to the emergency ward of
the nearest hospital and take along the pesticide label
and telephone number of the nearest poison control center.

First-aid treatment is extremely important, regardless of the time that may elapse before medical treatment
is available. The first-aid treatment received during the
first 2 to 3 minutes following a poisoning accident may
well spell the difference between life and death.

IMMEDIATE ACTION IS NECESSARY

FIRST - See that the victim is breathing; if not, give
        artificial respiration.
SECOND- Decontaminate him immediately, i.e., wash him off
        thoroughly. Speed is essential!
THIRD - Call your physician.

NOTE: Do not substitute first aid for professional treat-
        ment. First aid is only to relieve the patient
        before medical help is reached.

IN A POISONING EMERGENCY
CALL THE NEAREST LOCAL POISON CONTROL CENTER
OR
THE NEAREST REGIONAL POISON CONTROL CENTER
LISTED ON PAGE 219

GENERAL

1. Give mouth-to-mouth artificial respiration if breathing
   has stopped or is labored.
2. Stop exposure to the poison and if poison is on skin
   cleanse the person, including hair and fingernails. If
   swallowed, induce vomiting as directed (see action on
   swallowed poisons).
3. Save the pesticide container and material in it if any
   remains; get readable label or name of chemical(s) for
   the physician. If the poison is not known, save a
   sample of the vomitus.

SPECIFIC

Poison on Skin

1. Drench skin and clothing with water (shower, hose, faucet).
2. Remove clothing.
3. Cleanse skin and hair thoroughly with soap and water; rapidity in washing is most important in reducing extent of injury.
4. Dry and wrap in blanket.
5. Avoid use of ointments, greases, powders, and other drugs in first-aid treatment of burns.

Swallowed Poisons

1. CALL A PHYSICIAN IMMEDIATELY.
2. Do not induce vomiting if:
   a. Patient is in a coma or unconscious
   b. Patient is in convulsions.
   c. Patient has swallowed petroleum products (that is, kerosene, gasoline, lighter fluid).
   d. Patient has swallowed a corrosive poison (strong acid or alkaline products) - symptoms: severe pain, burning sensation in mouth and throat.
3. If the patient can swallow after ingesting a corrosive poison, give the following substances by mouth. A corrosive substance is any material that in contact with living tissue will cause destruction of tissue by chemical action, such as lye, acids, Lysol, etc.
   For acids and alkalis: Milk or water; for patients 1 to 5 years old, 2 to 4 ounces; for patients 5 years and older, up to 8 ounces.
4. If a non-corrosive substance has been swallowed, induce vomiting, if possible, EXCEPT if patient is unconscious, in convulsions, or has swallowed a petroleum product.
   a. Induce vomiting with SYRUP OF IPECAC. For persons over 1 year of age, including adults, give 15 cc (1 tablespoon) of syrup of Ipecac followed by water. For children: 1 to 5 years old, give up to 6 ounces of water, and up to 8 ounces for persons over 5 years of age. Ipecac requires about 20 minutes to produce vomiting.
   b. When retching and vomiting begin, place patient face down with head lowered, thus preventing vomitus from entering the lungs and causing further damage. Do not let him lie on his back.
   c. Do not waste excessive time in inducing vomiting if the hospital is a long distance away. It is better to spend the time getting the patient to the hospital where drugs can be administered to induce vomiting and/or stomach pumps are available.
   d. Clean vomitus from person. Collect some in case physician needs it for chemical tests.

Poison in Eye

1. Hold eyelids open, wash eyes with gentle stream of clean running water immediately. Use copius amounts. Delay of a few seconds greatly increases extent of injury.
2. Continue washing for 15 minutes or more.
3. Do not use chemicals or drugs in wash water. They may increase the extent of injury.

THE FIRST-AID TREATMENT RECEIVED DURING THE FIRST 2 TO 3 MINUTES FOLLOWING A POISONING ACCIDENT MAY MAKE THE DIF-FERENCE BETWEEN LIFE AND DEATH!

125 Review Questions That You Should
Be Able to Answer on Completion of
This Manual. (Number after question
indicates page where answer can be found.)

1. Name 8 pests of agriculture and 5 of suburbia. (1)
2. What is the annual pesticide usage per person in the
   U. S.? (1)
3. List 4 diseases of man or plants that changed history.
   (3)
4. What prevents us from returning to the agricultural
   practices of the "good old days?" (Preface)
5. What would be the outcome of withholding all pesticide
   use in the U. S. for one year? (5)
6. Why are pesticides essential today in agriculture and
   public health? (5)
7. What does the suffix -cide mean? (8)
8. By volume which 3 classes of pesticides are in the
   greatest use? (9)
9. Name the pesticide class that would be used to control
   mites? (10)
10. What pesticides are used for drying green plants? (11)
11. What is the chemical difference between organic and
    inorganic pesticides? (15)
12. Distinguish between molecular and structural formulas.
    (17)
13. What are the three types of formulas used to ullustrate
    chemical compounds? (17-19)
14. Of what value is the common name of a pesticide? (21)
15. List 5 reasons why pesticides are formulated. (23)
16. Which pesticide formulation is used most commonly? (27)
17. Which formulation group is undergoing the greatest
    changes? (30)
18. What is the maximum volume applied per acre as the ULV
    formulation? (31)
19. Which pesticide formulation is deposited least on target
    and drifts most? (33)
20. What is the meaning of 40 mesh as it refers to granular
    pesticides? (33)
21. Give two reasons why pesticides would be formulated in
    slow-release microcapsules. (36)
22. Why are the organochlorine insecticides being phased out
    by the EPA? (41)
23. What are the 4 classes of organochlorine insecticides?(52)
24. Generally what is the mode of action for the organo-
    chlorine insecticides? (52)
25. Give 4 synonyms for the organophosphate insecticides. (53)
26. What is the mode of action of the organophosphate
    insecticides? (54)
27. What are the functions of acetylcholine and cholin-
    esterase in the nervous system? (54)
28. What are the 3 sub-classes of organophosphate insecti-
    cides? (56,60,64)

244   *FINAL EXAM* (Continued)

29. Besides organophosphates, what other class of insecti-
    cides inhibits cholinesterase? (68)
30. How would you class the carbamate insecticides, persis-
    tent or non-persistent? (70,71)
31. What is the mode of action for the new formamidine
    insecticides? (74)
32. Name 4 botanical insecticides. (83)
33. Why are we just now getting around to using synthetic
    pyrethroids in agriculture? (84,85)
34. The synthetic pyrethroids have the same mode of action
    as pyrethrum. What is it? (85)
35. How do synergists improve the action of insecticides?
    (87,88)
36. Why is sulfur a good insecticide/miticide to include in
    integrated pest management programs? (90)
37. Why is the EPA cautious in registering the microbial
    insecticides? (95)
38. Give examples of the 3 generations of insecticides. (98)
39. List 5 non-agricultural uses for herbicides. (103)
40. Which category of herbicides is generally used for
    annual weeds and which for perennial? (104)
41. What are the 3 growth-timing categories for herbicides?
    (105)
42. Why is EPA gradually phasing-out the inorganic herbi-
    cides? (107)
43. The actions of phenoxy herbicides resemble those of
    which growth hormones? (110)
44. What happens when an herbicide inhibits the Hill reac-
    tion in plants? (112)
45. Which is probably the heaviest used group of herbicides
    in agriculture? (114)
46. Why are some plants tolerant to the triazine herbicides?
    (125)
47. The nitrophenols are used as insecticides, fungicides,
    herbicides, and defoliants. How do they react with
    these living systems? (132)
48. Generally which plant diseases are the most difficult
    to control with fungicides? (140)
49. Name 6 different organisms that cause plant diseases.
    (140)
50. Explain the delicate problem of controlling fungal
    diseases on plants. (140)
51. What is the difference between protectant, therapeutant,
    and eradicant fungicides? (141,142)
52. How do "fixed" copper fungicides work when they are
    practically insoluble? (144)
53. Why were the mercurial fungicides phased-out by the EPA?
    (146)
54. What are the advantages of the new synthetic fungicides
    over the old? (146,147)
55. Why are chelates important in plant metabolism? (148)
56. What is the major beneficial quality of the systemic
    fungicides? (153)

57. Why have certain disease organisms developed resistance toward the synthetic but not the heavy metal fungicides? (160)
58. What group of fungicides is effective against the bacterial diseases? (163)
59. Which fungicide group is not phytotoxic, trialkyl- or triaryl-tin? (166)
60. Why are highly-penetrating chemicals required as nematicides? (169)
61. What are the 4 chemical classes of nematicides? (169)
62. What is the mode of action of the carbamate and organophosphate nematicides? (172,175)
63. Why are the coumarin redenticides the safest and most commonly used? (177)
64. What is the mode of action of the coumarin rodenticides? (177)
65. What is the major difference between the first coumarin rodenticides and those developed recently? (179)
66. Rats cannot perform one important function that all other mammals can. What is it? (181)
67. Why is Compound 1080 limited to use by trained personnel? (182)
68. Name 6 uses of plant growth regulators. (184)
69. Name 4 of the 6 classes of plant growth regulators. (185)
70. How do plants generally respond to gibberellins? (186)
71. The growth inhibitors and retardants are antagonistic to which regulators? (190)
72. What are the benefits of using defoliants before harvest? (194)
73. What 2 conditions are necessary for defoliation? (194)
74. The organophosphate defoliants cause what changes that result in defoliation? (195)
75. What is the prime advantage of using a desiccant rather than a defoliant? (198)
76. What was the first federal law to control pesticides? (201)
77. What was the first federal law requiring accurate labeling of pesticides? (202)
78. The Miller Amendment to the Food and Drug Law controls what aspect of pesticides? (202)
79. The Delaney Amendment to the Food and Drug Law pertains specifically to what aspect of pesticides as food additives? (202,203)
80. What are carcinogens? (203)
81. What caution words are now required on all pesticide labels? (203)
82. What agency is responsible for enforcing FEPCA? (205)
83. What agency is responsible for enforcing pesticide tolerances in foods and feeds? (203)
84. All pesticides are classed into what two categories? (204)
85. Who is permitted to use restricted use pesticides? (204)
86. What is the most important pesticide law of this century? (204)

87. How many categories will be available in most states for the certification of commercial applicators? (205)
88. What general standards of knowledge are required for certification of all commercial applicators by FEPCA? (205)
89. How can a pesticide be used in a manner inconsistent with the label? (207)
90. In the misuse of pesticides who has primary enforcement responsibility? (207)
91. What is Special Review and what purpose does it serve? (208)
92. What is the Acceptable Daily Intake (ADI) of a pesticide? (210)
93. What local agency can provide antidote and treatment information for pesticide poisoning incidents? (218)
94. Whom would you consult regarding a certain pest and the pesticides recommended for its control? (213)
95. Why keep records of pesticide usage? (215)
96. Where can information be obtained on toxic waste and pesticide disposal sites? (215,216)
97. Where do you dispose of old or unwanted pesticides? (215)
98. For a serious pesticide spill, would you get help from CHEMTREC, your Local Poison Control Center, or your Regional Poison Control Center? (218)
99. What are the two types of risks associated with the use of pesticides? (221)
100. What is the meaning of LD_{50}? (224)
101. LD_{50}s are expressed in what terms? (224)
102. Of the 2 different types of pesticide toxicity data, which is the most useful to the applicator? (225)
103. What laboratory animals are used in obtaining pesticide toxicity data? (224)
104. Which EPA category of pesticides is most toxic? (227)
105. Of the insecticides, which class is the most toxic dermally? (229,232)
106. Which group of solvents used in pesticide formulations is most toxic to man? (232)
107. Generally which formulations of any pesticide are the most toxic dermally? (232)
108. Do any of the materials you work with appear in the Dangerous or Most Dangerous categories of Panel N, based on acute dermal hazard? (233,234)
109. Reentry intervals were established to prevent unnecessary dermal or inhalation exposure by field workers? (235)
110. Why are some formulations of a pesticide classed as restricted-use and others are not? (236)
111. Which insecticide class is growing most rapidly in numbers? (84)
112. What is the future for the halogenated fumigants and nematicides? (93,170)
113. The biorational pesticides may become the wave of the future. Why? (98)

114. What is the newest concept in weed control with chemicals? (135)
115. Why are applications of fungicides generally more frequent than insecticides to a given crop? (140,141)
116. In which group are the newest, fastest growing number of fungicides? (158)
117. Nematicides as a class of pesticides are in somewhat serious trouble. Why? (175)
118. Which category of pesticide must have DANGER - POISON on its label? (228)
119. In poisoning fatalities across the nation, which group of chemicals is responsible for the largest number? (222)
120. Why are pesticides safe to use if the label is followed when applying them? (223,224)
121. What route of exposure is most common with persons working with pesticides regularly? (225)
122. What class of pesticides contains a commercially available nematode, an antibiotic, and a protozoan? (97)
123. Can you make a toxicity comparison of the insect growth regulators with the synthetic pyrethroids to warm blooded animals? (99,233)
124. How safe are the "natural" insecticides, the botanicals, to man and his domestic animals? (83)
125. Now that you have completed the book, you should be able to designate which class of pesticides generally contains the most toxic compounds. (229,230,231,233)

GLOSSARY

Abscission - Process by which a leaf or other part is separated from the plant.

Acarcide (miticide) - An agent that destroys mites and ticks.

Acetylcholine (ACh) - Chemical transmitter of nerve and nerve-muscle impulses in animals.

Active ingredient (a.i.) - Chemicals in a product that are responsible for the pesticidal effect.

Acute toxicity - The toxicity of a material determined at the end of 24 hours; to cause injury or death from a single dose or exposure.

Adjuvant - An ingredient that improves the properties of a pesticide formulation. Includes wetting agents, spreaders, emulsifiers, dispersing agents, foam suppressants, penetrants, and correctives.

Adulterated pesticide - A pesticide that does not conform to the professed standard or quality as documented on its label or labeling.

Algicide - Chemical used to control algae and aquatic weeds.

Annual - Plant that completes its life cycle in one year, i.e., germinates from seed, produces seed, and dies in the same season.

Antagonism - Decreased activity arising from the effect of one chemical or another (opposite of synergism).

Antibiotic - Chemical substance produced by a microorganism and that is toxic to other microorganisms.

Anticoagulant - A chemical that prevents normal blood-clotting. The active ingredient in some rodenticides.

Antitranspirant - A chemical applied directly to a plant that reduces the rate of transpiration or water loss by the plant.

Aromatics - Solvents containing benzene or compounds derived from benzene.

Atropine (atropine sulfate) - An antidote used to treat organophosphate and carbamate poisoning.

Attractant, insect - A substance that lures insects to trap or poison-bait stations. Usually classed as food, oviposition, and sex attractants.

Auxin - Substance found in plants that stimulates cell growth in plant tissues.

Avicide - Lethal agent used to destroy birds but also refers to materials used for repelling birds.

Bactericide - Any bacteria-killing chemical.

Bentonite - A colloidal native clay (hydrated aluminum silicate) that has the property of forming viscous suspensions(gels) with water; used as a carrier in dusts; has good sticking qualities.

GLOSSARY (Continued)

Biennial - Plant that completes its growth in 2 years. The
    first year it produces leaves and stores food; the
    second year it produces fruit and seeds.
Biocide - A nondescript word found most frequently in sen-
    sational journalism, usually used as a synonym for
    pesticide, suggesting a chemical substance that kills
    living things or tissue.
Biological control agent - Any biological agent that
    adversely affects pest species.
Biomagnification - The increase in concentration of a
    pollutant in animals as related to their position in
    a food chain, usually referring to the persistent,
    organochlorine insecticides and their metabolites.
Biorational pesticides - Biological pesticides include pest
    control agents and chemical analogues of naturally
    occurring biochemicals (pheromones, insect growth
    regulators, etc.)

Biotic insecticide - Usually microorganisms known as insect
    pathogens that are applied in the same manner as con-
    ventional insecticides to control pest species.
Botanical pesticide - A pesticide produced from naturally
    occurring chemicals found in some plants. Examples
    are nicotine, pyrethrum, strychnine, and rotenone.
Broad-spectrum insecticide - Nonselective, having about
    the same toxicity to most insects.
Carbamate insecticide - One of a class of insecticides
    derived from carbamic acid.
Carcinogen - A substance that causes cancer in animal
    tissue.
Carrier - An inert material that serves as a diluent or
    vehicle for the active ingredient or toxicant.
Causal organism - The organism (pathogen) that produces
    a given disease.
Certified applicator - Person qualified to apply or super-
    vise application of restricted-use pesticides as
    defined by the EPA.
Chelating agent - Certain organic chemicals (i.e., ethyl-
    enediaminetetraacetic acid) that combine with metal
    to form soluble chelates and prevent conversion to
    insoluble compounds.
Chemical name - Scientific name of the active ingredient(s)
    found in the formulated product. The name is derived
    from the chemical structure of the active ingredient.
Chemosterilant - Chemical compounds that cause steriliza-
    tion or prevent effective reproduction.
Chemotherapy - Treatment of a diseased organism with chem-
    icals to destroy or inactivate a pathogen without
    seriously affecting the host.
Cholinesterase (ChE) - An enzyme of the body necessary for
    proper nerve function that is inhibited or damaged by
    organophosphate or carbamate insecticides taken into
    the body by any route.

GLOSSARY (Continued)

Chronic toxicity - The toxicity of a material determined
    beyond 24 hours and usually after several weeks of
    exposure.
Compatible - *Compatibility* - when two materials can be
    mixed together with neither affecting the action of
    the other.
Concentration -Content of a pesticide in a liquid or dust;
    for example, pounds/gallon or percent by weight.
Contact herbicide - Herbicide that causes localized injury
    to plant tissue where it touches.
Cumulative pesticides - Those chemicals that tend to
    accumulate or build up in the tissues of animals or in
    the environment (soil, water).
Curative pesticide - A pesticide that can inhibit or
    eradicate a disease-causing organism after it has
    become established in the plant or animal.
Cutaneous toxicity - Same as dermal toxicity.
Days-to-harvest - The least number of days between the
    last pesticide application and the harvest date, as
    set by law. Same as "harvest intervals".
Deflocculating agent - Material added to a spray prepara-
    tion to prevent aggregation or sedimentation of the
    solid particles.
Defoliant - A chemical that initiates abscission.
Dermal toxicity - Toxicity of a material as tested on the
    skin, usually on the shaved belly of a rabbit; the
    property of a pesticide to poison an animal or human
    when absorbed through the skin.
Desiccant - A chemical that induces rapid desiccation of
    a leaf or plant part.
Diatomaceous earth - A white powder prepared from naturally
    occurring deposits formed by the silicified skeletons
    of diatoms, used as a diluent in dust formulations and
    also as an insecticide.
Diluent - Component of a dust or spray that dilutes the
    active ingredient.
DNA - Deoxyribonucleic acid.
Dormant spray - Chemical applied in winter or very early
    spring before treated plants have started active
    growth.
Dose, dosage - Same as rate. The amount of toxicant given
    or applied per unit of plant, animal, or surface.
EC_{50} - The median effective concentration (ppm or ppb) of
    the toxicant in the environment (usually water) that
    produces a designated effect in 50 percent of the test
    organisms exposed.
Ecdysone - Hormone secreted by insects essential to the
    process of molting from one stage to the next.
Ecology - Derived from the Greek *oikos*,"house or place to
    live." A branch of biology concerned with organisms
    and their relation to the environment.

Economic injury level - The lowest population density of a
    pest that will cause economic damage (loss).
Economic threshold - The density of a pest at which control
    measures should be initiated to prevent an increasing
    pest population from reaching the economic injury
    level.
Ecosystem - The interacting system of all the living organ-
    isms of an area and their nonliving environment.
ED_{50} - The median effective dose, expressed as mg/kg of
    body weight, which produces a designated effect in
    50 percent of the test organisms exposed.
Emulsifiable concentrate - Concentrated pesticide formula-
    tion containing organic solvent and emulsifier to
    facilitate emulsification with water.
Emulsifier - Surface active substances used to stabilize
    suspensions of one liquid in another; for example,
    oil in water.
Emulsion - Suspension of miniscule droplets of one liquid
    in another.
Encapsulated formulation - Pesticide enclosed in capsules
    (or beads) of thin polyvinyl or other material, to
    control the rate of release of the chemical and extend
    the period of diffusion.
Environment - All the organic and inorganic features that
    surround and affect a particular organism or group
    of organisms.
EPA Establishment Number - A number assigned to each pesti-
    cide production plant by EPA. The number indicates
    the plant at which the pesticide product was produced
    and must appear on all labels of that product.
EPA Registration Number - A number assigned to a pesticide
    product by EPA when the product is registered by the
    manufacturer of his designated agent. The number must
    appear on all labels for a particular product.
Eradicant - Applies to fungicides in which a chemical is
    used to eliminate a pathogen from its host or environ-
    ment.
Exterminate - Often used to imply the complete extinction
    of a species over a large continuous area such as an
    island or a continent.
FEPCA - The Federal Environmental Pesticide Control Act of
    1972.
FIFRA - The Federal Insecticide, Fungicide and Rodenticide
    Act of 1947.
Fixed coppers - Insoluble copper fungicides where the cop-
    per is in a combined form. Usually finely divided,
    relatively insoluble powders.
Foliar application - Application of a pesticide to the
    leaves or foliage of plants.
Food chain - Sequence of species within a community, each
    member of which serves as food for the species next
    higher in the chain.

Formamidine insecticide - Insecticide class with a mode of
    action highly effective against insect eggs and mites.
Formulation - Way in which basic pesticide is prepared for
    practical use. Includes preparation as wettable pow-
    der, granular, emulsifiable concentrate, etc.
Full-coverage spray - Applied thoroughly over the crop to
    a point of runoff or drip.
Fumigant - A volatile material that forms vapors that
    destroy insects, pathogens, and other pests.
Fungicide - A chemical that kills fungi.
Fungistatic - Action of a chemical that inhibits the ger-
    mination of fungus spores while in contact.
Gallonage - Number of gallons of finished spray mix applied
    per acre, tree, hectare, square mile, or other unit.
General-use pesticide - A pesticide that can be purchased
    and used by the general public without undue hazard
    to the applicator and environment as long as the in-
    structions on the label are followed carefully. (See
    Restricted-use pesticide).
Granular - A dry formulation of pesticide and other com-
    ponents in discrete particles generally less than 10
    cubic millimeters in size.
Growth regulator - Organic substance effective in minute
    amounts for controlling or modifying (plant or insect)
    growth processes.
Harvest intervals - Period between last application of a
    pesticide to a crop and the harvest as permitted by
    law.
Hormone - A product of living cells that circulates in the
    animal or plant fluids and that produces a specific
    effect on cell activity remote from its point or
    origin.
Hydrolysis - Chemical process of (in this case) pesticide
    breakdown or decomposition involving a splitting of
    the molecule and addition of a water molecule.
Inert ingredient - Any substance in a pesticide product
    having no pesticidal action. This does not neces-
    sarily mean that it is an inactive ingredient.
Insect-growth regulator (IGR) - Chemical substance that
    disrupts the action of insect hormones controlling
    molting, maturity from pupal stage to adult, and
    others.
Integrated pest management - A management system that uses
    all suitable techniques and methods in as compatible
    a manner as possible to maintain pest populations at
    levels below those causing economic injury.
Invert emulsion - One in which the water is dispersed in
    oil rather than oil in water. Usually a thick mix-
    ture like salad dressing results.
IR-4 - Interregional Research Project No. 4. A national
    program devoted to the registration of pesticides
    for minor crops, headquartered at Rutgers University
    in New Jersey.

Larvicide - More commonly refers to chemicals used for con-
    trolling mosquito larvae (wiggle tails), but also to
    chemicals for controlling caterpillars on crops.
Layby - Applied with or after the last cultivation of a
    crop, usually herbicide.
LC_{50} - The median lethal concentration, the concentration
    that kills 50 percent of the test organisms, expressed
    as milligrans (mg) or cubic centimeters (cc), if liquid,
    per animal. It is also the concentration expressed as
    parts per million (ppm) or parts per billion (ppb) in
    the environment (usually water) that kills 50 percent
    of the test organisms exposed.
LD_{50} - A lethal dose for 50 percent of the test organisms.
    The dose of toxicant producing 50 percent mortality
    in a population. A value used in presenting mam-
    malian toxicity, usually oral toxicity, expressed as
    milligrams of toxicant per kilogram of body weight
    (mg/kg).
mg/kg (milligrams per kilogram) - Used to designate the
    amount of toxicant required per kilogram of body
    weight of test organism to produce a designated effect,
    usually the amount necessary to kill 50 percent of the
    test animals.
Microbial insecticide - A microorganism applied in the same
    way as conventional insecticides to control an exist-
    ing population.
Miscible liquids - Two or more liquids capable of being
    mixed in any proportions and of remaining mixed under
    normal conditions.
M.L.D. - Median lethal dose (LD_{50}).
Molluscicide - A chemical used to kill or control snails
    and slugs.
Minimum tillage - Practices which utilize minimum cultiva-
    tion for seedbed preparation and may reduce labor and
    fuel costs; may also reduce damage to soil structure,
    e.g., erosion.
Mutagen - Substance causing genes in an organism to mutate
    or change.
Mycoplasma - A microorganism intermediate in size between
    viruses and bacteria possessing many virus-like prop-
    erties and not visible with a light microscope.
Negligible residue - A tolerance which is set on a food or
    feed crop permitting an ultra-small amount of pesti-
    cide at harvest as a result of indirect contact with
    the chemical.
Nematicide - Chemical used to kill nematodes.
Neurotoxicity - The ability to cause defects in nerve
    tissue.
Nonselective herbicide - A chemical that is generally toxic
    to plants without regard to species. Toxicity may be
    a function of dosage, method of application, etc.
Nuclear polyhedrosis virus (NPV) - A disease virus of in-
    sects, cultured commercially and sold as a biological
    insecticide.

Oral toxicity - Toxicity of a compound when given by mouth.
Usually expressed as number of milligrams of chemical
per kilogram of body weight of animal (white rat) when
given orally in a single dose that kills 50 percent of
the animals. The smaller the number, the greater the
toxicity.
Organochlorine insecticide - One of the many chlorinated
insecticides, e.g., DDT, dieldrin, chlordane, BHC,
Lindane, etc.
Organophosphate - Class of insecticides (also some herbi-
cides, defoliants and fungicides) derived from
phosphoric acid esters.
Organotins - A classification of miticides and fungicides
containing tin as the nucleus of the molecule.
Ovicide - A chemical that destroys an organism's eggs.
Pathogen - Any disease-producing organism or virus.
Penetrant - An additive or adjuvant which aids the pesticide to move through the outer surface of plant
tissues.
Perennial - Plants that continue to live from year to year.
Plants may be herbaceous or woody.
Pesticide - An "economic poison" defined in most state and
federal laws as any substance used for controlling,
preventing, destroying, repelling, or mitigating any
pest. Includes fungicides, herbicides, insecticides,
nematicides, rodenticides, desiccants, defoliants,
plant growth regulators, etc.
Pheromones - Highly potent insect sex attractants produced
by the insects. For some species, laboratory-synthe-
sized pheromones have been developed for trapping purposes.
Phytotoxic - Injurious to plants.
Piscicide - Chemical used to kill fish.
Plant Regulator (Growth regulator) - A chemical which
increases, decreases, or changes the normal growth or
reproduction of a plant.
Postemergence - After emergence of a specified weed or
crop.
ppb - Parts per billion (parts in 10^9 parts) is the number
of parts of toxicant per billion parts of the substance in question.
ppm - Parts per million (parts in 10^6 parts) is the number
of parts of toxicant per million parts of the substance in question. They may include residues in crops,
soil, water, or whole animals.
Predacide - Chemicals used to poison predators.
Preemergence - Applied prior to emergence of the specified
weed or planted crop.
Preplant application - Treatment applied on the soil surface before seeding or transplanting.
Preplant soil incorporated - Herbicide applied and tilled
into the soil before seeding or tansplanting.

Protectant - Fungicide applied to plant surface before
     pathogen attack to prevent penetration and subsequent
     infection.
Protopam chloride (2-pam) - An antidote for certain organo-
     phosphate pesticide poisoning, but not for carbamate
     poisoning.
Raw agricultural commodity - Any food in its raw and
     natural state, including fruits, vegetables, nuts,
     eggs, raw milk, and meats.
Rebuttable Presumption Against Registration (RPAR)
     See Special Review.
Reentry (intervals) - Waiting interval required by federal
     law between application of certain hazardous pesti-
     cides to crops and the entrance of workers into those
     crops without protective clothing.
Repellent (insects) - Substance used to repel ticks, chig-
     gers, gnats, flies, mosquitoes, and fleas.
Residual herbicide - An herbicide that persists in the soil
     and injures or kills germinating weed seedlings over
     a relatively short period of time (see persistent
     herbicide).
Residue - Trace of a pesticide and its metabolites remain-
     ing on and in a crop, soil, or water.
Resistance - Natural or genetic ability of an organism to
     tolerate the poisonous effects of a toxicant.
Restricted-use pesticide - One of several pesticides, des-
     ignated by the EPA, that can be applied only by certi-
     fied applicators, because of their inherent toxicity
     or potential hazard to the environment.
RNA - Ribonucleic acid.
Rodenticide - Pesticide applied as a bait, dust, or fumi-
     gant to destroy or repel rodents and other animals.
Ropewick applicator - A rope saturated with a foliage-
     applied translocated herbicide solution which is wiped
     across the surface of weed foliage. The rope utilizes
     forces of capillary attraction and conducts the herbi-
     cide from a reservoir.
RPAR - See Special Review.
Safener - Chemical that reduces the toxicity of another
     chemical.
Selective pesticide - One that, while killing the pest
     individuals, spares much or most of the other fauna or
     flora, including beneficial species, either through
     differential toxic action or through the manner in
     which the pesticide is used (formulation, dosage,
     timing, placement, etc.).
Senescence - Process or state of growing old.
Signal word - A required word that appears on every pesti-
     cide label to denote the relative toxicity of the
     product. The signal words are either *Danger - Poison*
     for highly toxic compounds, *Warning* for moderately
     toxic, or *Caution* for slightly toxic.
Silvicide - A term applied to herbicides used to control
     undesirable brush and trees, as in wooded areas.

Slimicide - Chemicals used to prevent slimy growth, as in
    wood-pulping processes for manufacture of paper and
    paperboard.
Slurry - Thin, watery mixture, such as liquid mud, cement,
    etc. Fungicides and some insecticides are applied to
    seeds as slurries to produce thick coating and reduce
    dustiness.
Soil incorporation - Mechanical mixing of herbicide with
    the soil.
Special Review (formerly Rebuttable Presumption Against
    Registration (RPAR)). A regulatory investigation pro-
    cess used by EPA when a pesticide shows potentially
    dangerous characteristics. At the conclusion of the
    investigation, the chemical is (1) returned to full
    registration, (2) some or all uses become restricted,
    (3) announcement of intent to cancel or suspend some
    or all uses, or (4) a combination of these.
Spreader - Ingredient added to spray mixture to improve
    contact between pesticide and plant surface.
Sticker - Ingredient added to spray or dust to improve its
    adherence to plants.
Stomach poison - A pesticide that must be eaten by an
    insect or other animal in order to kill or control the
    animal.
Stupefacient or soporific - Drug used as a pesticide to
    cause birds to enter a state of stupor so they can be
    captured and removed, or to frighten other birds away
    from the area.
Surfactant - Ingredient that aids or enhances the surface-
    modifying properties of a pesticide formulation
    (wetting agent, emulsifier, or spreader).
Suspension - Finely divided solid particles dispersed in
    a liquid.
Synergism - Increased activity resulting from the effect of
    one chemical on another.
Synthesize - Production of a compound by joining various
    elements or simpler compounds.
Systemic - Compound that is absorbed and translocated
    throughout the plant or animal.
Tank mix - Mixture of two or more pesticides in the spray
    tank at time of application, provided the current
    labeling does not prohibit this practice.
Target - The plants, animals, structures, areas, or pests
    to be treated with a pesticide application.
Teratogenic - Substance that causes physical birth defects
    in the offspring following exposure of the pregnant
    female.
Threshold limit value (TLV) - The maximum air concentration
    of a chemical, expressed as milligrams per cubic meter
    (mg/m^3), in which workers may perform their duties
    8 hours a day, 40 hours per week, with no adverse
    health effects, as established by the American Con-
    ference of Governmental Hygienists.

TLV - See threshold limit value.

Tolerance - Amount of pesticide residue permitted by federal regulation to remain on or in a crop at harvest. Expressed as parts per million (ppm).

Toxicant - A poisonous substance such as the active ingredient in pesticide formulations that can injure or kill plants, animals, or mircroorganisms.

Toxic Substances Control Act (TSCA) - This Act (PL 94-469) became effective January 1, 1977, and gives broad authority to EPA to obtain information from industry on the production, use, health effects, and other matters concerning chemical substances and mixtures. Chemicals already regulated by other laws are exempt from this Act (pesticides, tobacco, ammunition, food, food additives, drugs, cosmetics and nuclear material).

Toxin - A naturally occurring poison produced by plants, animals, or microorganisms; for example, the poison produced by the black widow spider, the venom produced by snakes, and the botulism toxin.

Toxogonin^R (bis-(4-hydroxy-iminomethyl-pyridinum-(1)-methyl) ether dichloride) - An antidote used in the treatment of organophosphate pesticide poisoning. Its action is identical to that of 2-PAM, and is not to be used in the treatment of carbamate poisoning. Toxogonin^R is manufactured by Merck (Germany) and is neither sold nor used in the United States.

Trade name (trademark name, proprietary name, brand name) - Name given a product by its manufacturer or formulator, distinguishing it as being produced or sold exclusively by that company.

Translocation - Transfer of food or other materials such as herbicides from one plant part to another.

Trivial name - Name in general or commonplace usage; for example, nicotine.

Ultra-low volume (ULV) - Sprays that are applied at 0.5 gallon or less per acre or sprays applied as the undiluted formulation.

Vector - An organism, as an insect, that transmits pathogens to plants or animals.

Viricide - A substance that inactivates a virus completely and permanently.

Virustatic - Prevents the multiplication of a virus.

Volatilize - To vaporize.

Wettable powder - Pesticide formulation of toxicant mixed with inert dust and a wetting agent that mixes readily with water and forms a short-term suspension (requires tank agitation).

Wetting agent - Compound that causes spray solutions to contact plant surfaces more thoroughly.

Winter annual - Plant that starts germination in the fall, lives over winter, and completes its growth, including seed production, the following season.

BIBLIOGRAPHY

Anonymous. 1980. Pesticides: $6 Billion by 1990. Chemical Week.
    May 7.
Ashton, F. M., and A. S. Crafts. 1981. Mode of Action of Herbicides.
    2nd edition. John Wiley & Sons, New York, NY. 525 p.
Aspelin, A. L., and G. L. Ballard. 1984. Pesticide Industry Sales and
    Usage 1983 Market Estimates. Economic Analysis Br., Benefits and
    Field Studies Div., Office of Pesticide Programs, EPA, Washington,
    D. C. 20460. August, 9 p.
Burges, H. A., and N. W. Hussey (Eds). 1971. Microbial Control of
    Insects and Mites. Academic Press, New York, NY. 861 p.
Corbett, J. R. 1974. The Biochemical Mode of Action of Pesticides.
    Academic Press, New York, NY. 330 p.
Council for Agricultural Science and Technology. 1980. Organic and
    Conventional Farming Campared. CAST Report No. 84. Chairman,
    S. R. Aldrich. Iowa State University, Ames, Iowa 50011.
Delp, C. J. 1980. Coping with resistance to plant disease control
    agents. Plant Disease. 64(7):652-657. (July)
Delvo, Herman, and Michael Hanthorn. 1983. Pesticides. In Inputs--
    Outlook & Situation. U.S.D.A. Economic Research Service. pp.
    4-12. (October)
Erwin, D. C. 1973. Systemic Fungicides: Disease Control, Trans-
    location, and Mode of Action. Ann. Rev. Phytopathology.
    11:389-422.
Fowler, D. Lee, and J. N. Mahan. 1980. The Pesticide Review 1978.
    U.S.D.A., Agr. Stabilization & Conserv. Service, Washington,
    D. C. 42 p.
Halder, M. R., and R. S. Shadbolt. 1975. Novel 4-hydroxy-coumarin
    anticoagulants active against resistant rats. Nature, 253:275-77.
Hayes, W. J., Jr. 1975. Toxicology of Pesticides. Williams and
    Wilkins Co., Baltimore, MD. 580 p.
Holan, G. 1969. New halocyclopropane insecticides and the mode of
    action of DDT. Nature (London) 221:1025.
Hollingworth, R. M. 1976. The Biochemical and Physiological Basis of
    Selective Toxicity. In Insecticide Biochemistry and Physiology
    (C. F. Wilkinson, ed.), p. 431-506, Plenum Press, New York, NY.
Hollingworth, R. M. and L. L. Murdock. 1980. Formamidine Pesticides:
    Octopamine-like Actions in a Firefly. Science, 208, 74-76.
Kohn, G. K. (Ed.). 1974. Mechanism of Pesticide Action. American
    Chemical Society, Washington, D. C. 180 p.
Lukens, R. J. 1971. Chemistry of Fungicidal Action. Molecular
    Biology, Biochemistry and Biophysics Series, No. 10. Springer
    Verlag, New York, NY. 136 p.
Matsumura, Fumio. 1975. Toxicology of Insecticides. Plenum Press,
    New York, NY. 503 p.

Meister, R. T. (Ed.). 1985. Farm Chemicals Handbook. Meister
    Publishing Co., Willoughby, OH.
Melnikov, N. N. 1971. Chemistry of Pesticides. Springer Verlag,
    New York, NY. 480 p.
Menn, J. J., and M. Beroza (Eds.). 1972. Insect Juvenile Hormones,
    Chemistry and Action. Academic Press, New York, NY. 341 p.
Saleh, M. A., and J. E. Casida, 1978. Reductive Dechlorination of
    the Toxaphene Component 2,2,5-endo,6-exo,8,9,10-heptachlorobor-
    nane in Various Chemical, Photochemical, and Metabolic Systems.
    J. Agric. Food Chem. 26(3):583-590.
Staal, G. B. 1975. Insect Growth Regulators With Juvenile Hormone
    Activity. Ann. Rev. Entomology 20:417-460.
Storck, W. J. 1980. Pesticide Profits Belie Mature Market Status.
    Chem. Engr. News. April 28: 10-13.
Thomson, W. T. 1983. Agricultural Chemicals. Book II: Herbicides.
    Thomson Publications, Fresno, CA. 285 p.
Thomson, W. T. 1984. Agricultural Chemicals. Book IV: Fungicides.
    Thomson Publications, Fresno, CA. 181 p.
Timm, R. M. (Ed.). 1984. Prevention and Control of Wildlife Damage.
    Cooperative Extension Service, Univ. Nebraska, Lincoln. 550 p.
Torgeson, E. Y. (Ed.). 1967. Fungicides, An Advanced Treatise.
    Vol. 1: Agricultural and Industrial Applications, Environmental
    Interactions. Academic Press, New York, NY. 697 p.
Torgeson, E. Y. (Ed.). 1969. Fungicides, An Advanced Treatise.
    Vol. 2: Chemistry and Physiology. Academic Press, New York,
    NY. 742 p.
Van Valkenburg, W. 1973. Pesticide Formulations. Marcel Dekker,
    New York, NY. 481 p.
Ware, G. W. 1980. Complete Guide to Pest Control--With and Without
    Chemicals. Thomson Publications, Fresno, CA. 290 p.
Ware, G. W. 1983. Pesticides--Theory and Application. W. H.
    Freeman and Co., New York, NY. 308 p.
Weaver, R. J. 1972. Plant Growth Substances in Agriculture.
    W. H. Freeman and Co., New York, NY. 594 p.
Weed Science Society of America. 1983. Herbicide Handbook. 5th ed.
    Champaign, IL. 515 p.
Wilkinson, D. F. (Ed.). 1976. Insecticide Biochemistry and
    Physiology. Plenum Press, New York, NY. 768 p.
Worthing, C. R. (Ed.). 1983. The Pesticide Manual--A World Compendium.
    7th ed. British Crop Protection Council, 144-150 London Road,
    Croydon CRO2TD, Great Britain.